STABILITY THEORY
OF DIFFERENTIAL EQUATIONS

STABILITY THEORY
OF DIFFERENTIAL EQUATIONS

RICHARD BELLMAN

University of Southern California

DOVER PUBLICATIONS, INC.

New York

Published in Canada by General Publishing Company, Ltd., 30 Lesmill Road, Don Mills, Toronto, Ontario.
Published in the United Kingdom by Constable and Company, Ltd., 10 Orange Street, London WC 2.

This Dover edition, first published in 1969, is an unabridged and unaltered republication of the work originally published in 1953 by the McGraw-Hill Book Company, Inc.

Standard Book Number: 486-62210-X
Library of Congress Catalog Card Number: 79-84704

Manufactured in the United States of America
Dover Publications, Inc.
180 Varick Street
New York, N.Y. 10014

To Betty-Jo

PREFACE

The fundamental problem in the theory of differential equations is that of deducing the properties of the solutions of a given differential equation from the analytic form of the equation. Although occasionally and fortuitously this may be accomplished very simply by expressing the solution in terms of the elementary functions of analysis, in general the equations which appear in theoretical investigations, of both mathematical and physical origin, are not explicitly integrable. Rather they constitute the principal source of new transcendents whose properties may be determined only by a systematic and penetrating analysis of wide classes of equations.

We shall in this book consider only real solutions of real equations and the behavior of these solutions as the independent variable increases without limit. In problems of physical interest this variable is most frequently the time. The properties of the solution of greatest interest to us will be boundedness, asymptotic behavior, oscillation, and stability.

No attempt has been made to be encyclopedic and to catalogue every result that possibly falls within our scope. This is neither feasible nor desirable in an introductory volume. We have rather tried to unify the theory as much as possible by focusing attention on a small number of powerful techniques. It is for this reason that we have not scrupled occasionally to prove theorems several times over, employing varied approaches.

The arguments throughout are elementary, depending only upon the fundamental concepts of analysis. It is due to this fact that much of the work is quite difficult, since many of the results must be built up, block by block, tediously and laboriously, from the elemental notions, each new result requiring a new structure.

To preserve the elementary character of the work, we have omitted any discussion of periodic solutions of nonlinear differential equations, such as the famed equation of Van de Pol, since such discussion would require the use of quite advanced analytic and topological tools.

The plan of the book is as follows: In Chap. 1 we study the fundamental properties of linear systems, deriving the results which are essential for the later deeper study of linear and nonlinear systems. To that

end we introduce vector-matrix notation and study simple transformations of matrices which are of great utility in the theory of asymptotic
behavior. It has been intended that the account given of the small
amount of matrix theory required be self-contained. Unfortunately,
there is no single source to which we can refer the reader for any more
extensive treatment covering the topics that are needed.

In Chap. 2 we turn to the interesting and important question of determining the asymptotic behavior of solutions of equations whose coefficients are nearly constant, whose study was initiated by Dini and
Poincaré. After presenting a number of results which yield first-order
estimates, we consider the problem of obtaining approximations of
arbitrarily high order. This leads naturally to the concept of asymptotic
series, as introduced by Poincaré. Since the literature on this subject is
vast and much of the material quite complicated, we have, consistent
with our general aim, presented only one of the most important results in
order to enable the reader to taste the flavor of the general theory.

Chapter 2 also contains, in connection with the discussion of asymptotic
behavior, a discussion of the concept of stability, that much overburdened
word with an unstabilized definition.

Existence and uniqueness theorems for nonlinear systems constitute
the content of Chap. 3. These are given not so much for their own sake,
but because they furnish excellent motivation for exhibiting two powerful
methods, the method of successive approximations, already viewed in a
simpler setting in Chap. 1, and the method of approximating to differential equations by difference equations.

Next, in Chap. 4, we present the fundamental results of Poincaré and
Liapounoff concerning the stability of solutions of nonlinear systems.
In order to illustrate a variety of important techniques, we present
several proofs under conditions of varying degrees of restrictiveness.

Chapter 5 is dedicated to the study of real solutions of the polynomial
equation $P(t,u,du/dt) = 0$. Introducing the important concept of a
proper solution, a solution which remains finite for $t \geq t_o$, which is precisely the type of solution which is required in most physical investigations, we present the remarkable results of Borel and Hardy concerning
the asymptotic behavior of the real, proper solutions.

Chapter 6 presents results which are a combination of ingenuity and
special techniques. Just as the geometry of the triangle and of the circle
obstinately refuses to rest content as a set of corollaries of results valid
for general algebraic curves, and constantly furnishes its devotees new
theorems undreamt of in the broader discipline, so the study of the
equation $u'' + a(t)u = 0$ is replete with elegant and unexpected results
which are not to be derived from any general theory of the *nth*-order

linear system. We have attempted to present a sufficient number of devices (remembering that a device is a trick that works at least twice) to enable the diligent reader who has worked through the chapter to obtain new results and to read research papers.

Chapter 7 is devoted to a particular nonlinear equation of the form $u'' \pm t^m u^n = 0$. This equation came first into prominence in connection with the astrophysical researches of Emden. A number of results obtained by Emden in the usual half-intuitive, wholly ingenious fashion of the physicist were made precise by Fowler, who was then stimulated to continue and give a complete discussion of the proper solutions of this equation for all values of the parameters. This equation, and the closely related equation $u'' \pm e^{at} u^n = 0$, have become of increasing importance recently in nuclear physics, in connection with work of Fermi and Thomas.

The purpose of Chaps. 5 and 7 is not only to rescue from partial oblivion a number of extremely interesting results and techniques in the theory of differential equations, but also to illustrate the fact that non-linear differential equations are by no means the intransigent creatures they appear to be upon first frightened glance. Since modern physical theory is being driven more and more to rely upon nonlinear explanations of basic phenomena, we hope that the contents of these chapters may be some slight consolation for the lapse from the grace of superposition.

It is with great pleasure that I record at this time my deep appreciation to Solomon Lefschetz, who first guided my footsteps into the path of research in the theory of differential equations. His stimulating discussions and constant encouragement furnished continued sources of inspiration.

I should also like to acknowledge my gratitude to Mina Rees and the Office of Naval Research, who supported and encouraged my original work in the theory of differential equations. Finally, I should like to thank a number of friends who read various portions of the manuscript and made many helpful comments.

Richard Bellman

CONTENTS

CONTENTS   xiii

CHAPTER 1

PROPERTIES OF LINEAR SYSTEMS

1. Introduction. In this introductory chapter we shall consider the fundamental properties of solutions of the system of linear differential equations,

$$(1) \qquad \frac{dy_i}{dt} = \sum_{j=1}^{n} a_{ij}(t)y_j, \qquad i = 1, 2, \ldots, n$$

The independent variable t is to range over the interval $[0, \infty]$, and we shall assume that the coefficient functions $a_{ij}(t)$ are piecewise-continuous over any finite subinterval. Under this assumption, we may consider all integrals that appear to be Riemann integrals. For our purposes there is very little to be gained from the sophistication of the Lebesgue integral, and we prefer, in consequence, to keep our discussion on as elementary a level as possible.

We furthermore postulate that the coefficients are real functions. Occasionally, particularly in the discussion of linear systems with constant coefficients and with coefficients close to constant, we shall introduce complex solutions. For example, we may use (e^{it}, e^{-it}) as a basic set of solutions of $d^2u/dt^2 + u = 0$, rather than $(\cos t, \sin t)$. This is purely a matter of convenience, however, and we shall always be primarily interested in real solutions of real systems.

The only way to study the behavior of solutions of systems of linear algebraic equations or linear differential equations in any systematic fashion is to make use of the concepts of vectors and matrices. In this chapter we shall introduce these concepts and demonstrate the few results required for the theory of differential equations. No prior knowledge of vector or matrix theory will be assumed.

Exercise

Show that the nth-order linear equation

$$u^{(n)} + a_1(t)u^{(n-1)} + \cdots + a_n(t)u = 0$$

may be converted into a linear system of the type of (1) above by means of the substitutions $u = u_1$, $u' = u_2$, $\ldots$, $u^{(n-1)} = u_n$.

---

2. Vector-Matrix Notation. The column of n quantities,

$$(1) \qquad y = \begin{pmatrix} y_1 \\ y_2 \\ \cdot \\ \cdot \\ \cdot \\ y_n \end{pmatrix}$$

where the y_i are real or complex, will be called an n-dimensional *column vector*, and the symbol $(y_1, y_2, \ldots, y_n)$ will be called an n-dimensional *row vector*. If the y_i are functions of t, y will be called a vector function of t; otherwise, it will be called a constant vector. The quantity y_i is called the ith component of y. We shall, for the greater part, use column vectors.

The letters x, y, z, u, v, and w will be systematically reserved to represent vector functions, while a, b, c, and d will be used to represent constant vectors. As far as possible, u and v will be reserved to denote one-dimensional vectors which we call *scalars*, and $c_1, c_2, \ldots$ will be used to denote scalar constants.

Let us now consider various operations which may be performed upon vectors. The simplest is addition. The sum of two vectors x and y, written $x + y$, is defined to be the vector whose ith component is $x_i + y_i$. It follows from our definition that the operation of addition is commutative and associative. Using a limiting process, we are led to define the integral of $y = y(t)$ as the vector whose ith component is $\int y_i \, dt$, and we write $\int y \, dt$. The product of a scalar c_1 and a vector y is a vector $c_1 y$ whose ith component is $c_1 y_i$.

To measure the magnitude, or length, of a vector y, we introduce the scalar quantity

$$(2) \qquad \|y\| = \sum_{i=1}^{n} |y_i|$$

which we call the *norm* of y. It is readily verified that

$$(3) \qquad \|x + y\| \leq \|x\| + \|y\|, \qquad \text{(triangle inequality)}$$
$$\|c_1 y\| = |c_1| \|y\|$$
$$\|\int y \, dt\| \leq \int \|y\| \, dt$$

and that $\|y\| = 0$ if and only if every component of y is equal to zero. The advantage of this norm over the Euclidean norm $\left(\sum_{i=1}^{n} y_i^2 \right)^{1/2}$ lies in its simplicity.

Having defined vectors, we now introduce the concept of a square matrix, the only type of matrix we shall employ. The square array of numbers, real or complex,

$$(4) \qquad A = \begin{pmatrix} a_{11} & a_{12} & \cdots & a_{1n} \\ a_{21} & a_{22} & \cdots & a_{2n} \\ \cdot & \cdot & & \cdot \\ \cdot & \cdot & & \cdot \\ \cdot & \cdot & & \cdot \\ a_{n1} & a_{n2} & \cdots & a_{nn} \end{pmatrix} = (a_{ij})$$

will be called a *matrix* of order n. The quantity a_{ij} is called the ijth element of A. As before, A will be called a matrix function if its elements are functions of t, and otherwise a constant matrix. It will be said to be continuous in $[a,b]$ if its elements are continuous in this interval.

The sum of two matrices A and B is defined by

$$(5) \qquad A + B = (a_{ij} + b_{ij})$$

while the product is defined by

$$(6) \qquad AB = \left(\sum_{k=1}^{n} a_{ik} b_{kj} \right)$$

It is clear that addition is commutative and associative, but that multiplication, while always associative, is, in general, *not* commutative.

Exercise

1. Show that $A(B + C) = AB + AC$, and that

$$(B + C)A = BA + CA$$

A matrix of particular importance is the identity matrix

$$(7) \qquad I = \begin{pmatrix} 1 & 0 & \cdots & 0 \\ 0 & 1 & & 0 \\ \cdot & \cdot & & \cdot \\ \cdot & \cdot & & \cdot \\ \cdot & \cdot & & \cdot \\ 0 & 0 & \cdots & 1 \end{pmatrix}$$

For all A we have $AI = IA = A$.

The product of a scalar c_1 and a matrix A is the matrix $c_1 A = A c_1$ equal to $(c_1 a_{ij})$. The product of a column vector y by a matrix A is written Ay—note the order of the factors—and is defined to be the vector whose

ith component is $\displaystyle\sum_{j=1}^{n} a_{ij}y_j$. It is easily seen that ABy is unambiguous, being equal to $(AB)y = A(By)$.

The definitions of addition and multiplication, at first sight quite artificial, become reasonable and intuitive if we consider the matrix A to represent the linear transformation in n dimensions,

$$(8) \qquad x'_i = \sum_{j=1}^{n} a_{ij}x_j, \qquad i = 1, 2, \ldots, n$$

The resultant of the transformation represented by B followed by the transformation represented by A yields another transformation C, which we call AB. It is readily seen that this new definition of AB coincides with the one given above by (6). It is clear now, geometrically, why $AB \neq BA$ in general.

To measure the magnitude of A, we use the scalar quantity

$$(9) \qquad \|A\| = \sum_{i,j=1}^{n} |a_{ij}|$$

which we call the *norm* of A. It is easily seen that

$$(10) \qquad \begin{aligned} \|A + B\| &\leq \|A\| + \|B\| \\ \|AB\| &\leq \|A\|\,\|B\| \\ \|c_1 A\| &\leq |c_1|\,\|A\| \\ \|Ax\| &\leq \|A\|\,\|x\| \end{aligned}$$

As before, we define $\int A\, dt$ to be the matrix whose ijth element is $\int a_{ij}\, dt$.

Exercise

2. Show that $\left\| \int A\, dt \right\| \leq \int \|A\|\, dt$.

Having defined integration of vectors and matrices, we also define the inverse operation of differentiation in the expected fashion:

$$(11) \qquad \frac{dA}{dt} = \left(\frac{da_{ij}}{dt} \right)$$

$$\frac{dy}{dt} = \begin{pmatrix} \dfrac{dy_1}{dt} \\ \cdot \\ \cdot \\ \cdot \\ \dfrac{dy_n}{dt} \end{pmatrix}$$

In terms of this new notation, our fundamental linear system described in (1) of Sec. 1 takes the simple, elegant form,

$$(12) \qquad \frac{dy}{dt} = Ay$$

If we assign an initial value to y, it becomes

$$(13) \qquad \frac{dy}{dt} = Ay, \qquad y(0) = c$$

In the next section, we turn to the problem of determining whether or not (13) has a solution.

Before proceeding to this, we require a few more elementary facts about matrices. Associated with each matrix we have the scalar quantity, $|A|$, the determinant of A. If $|A| = 0$, we say that A is singular, otherwise nonsingular.

Exercise

3. Show that $|AB| = |A|\,|B|$.

The importance of this new concept lies in the fact that a nonsingular matrix A possesses a unique inverse, a matrix which we shall denote by A^{-1}. This matrix has the property that

$$(14) \qquad AA^{-1} = A^{-1}A = I$$

Exercises

4. Show that $A^{-1} = (\alpha_{ij}/|A|)$, where α_{ij} is the cofactor of a_{ji}. Show that $(A^{-1})^{-1} = A$, and that $(AB)^{-1} = B^{-1}A^{-1}$.

5. Show that

$$\frac{d}{dt}(AB) = \frac{dA}{dt}B + A\frac{dB}{dt}$$

$$\frac{d}{dt}(Ay) = \frac{dA}{dt}y + A\frac{dy}{dt}$$

$$\frac{d}{dt}A^{-1} = -A^{-1}\frac{dA}{dt}A^{-1}$$

We shall also require the notion of an infinite series of vectors or matrices. We define

$$(15) \qquad \sum_{m=1}^{\infty} A^{(m)} = \left(\sum_{m=1}^{\infty} a_{ij}^{(m)} \right)$$

$$\sum_{m=1}^{\infty} y^{(m)} = \begin{pmatrix} \sum_{m=1}^{\infty} y_1^{(m)} \\ \cdot \\ \cdot \\ \cdot \\ \sum_{m=1}^{\infty} y_n^{(m)} \end{pmatrix}$$

provided, of course, that the infinite series appearing on the right converge.

Exercise

6. Show that sufficient conditions for the convergence of $\sum_{n=1}^{\infty} A^{(n)}$ and $\sum_{m=1}^{\infty} y^{(m)}$ are that $\sum_{n=1}^{\infty} \|A^{(n)}\|$ and $\sum_{m=1}^{\infty} \|y^{(m)}\|$, respectively, converge.

---

3. Existence of Solutions of the Vector-Matrix Equation $dy/dt = A(t)y$. Our first theorem will be an existence and uniqueness theorem. The result is included in a later result concerning nonlinear systems, and the method is precisely the same as that used for the more general case. Nevertheless, we shall present the proof in all its details since it furnishes an excellent introduction, free of extraneous difficulties, to the techniques we shall employ in what follows.

Theorem 1. *Let $A(t)$ be continuous in the interval $[0,t_0]$. Then there exists a unique solution of*

$$(1) \qquad \frac{dy}{dt} = A(t)y, \qquad y(0) = c$$

in this interval.

Proof of Existence. Let us introduce a method which will be used frequently in what follows, the celebrated and fundamental method of successive approximations due to Picard.

Consider the sequence of (vector) functions defined inductively as follows:

(2) $$y_0 = c$$
$$\frac{dy_1}{dt} = A(t)y_0, \qquad y_1(0) = c,$$

$$\vdots$$

$$\frac{dy_{n+1}}{dt} = A(t)y_n, \qquad y_{n+1}(0) = c, \qquad n = 0, 1, 2, \ldots$$

This is equivalent to

(3) $$y_0 = c$$
$$y_{n+1} = c + \int_0^t A(t_1)y_n \, dt_1, \qquad n = 0, 1, 2, \ldots$$

We wish to show that the sequence of functions defined by (3) converges uniformly to a function $y(t)$ for $0 \leq t \leq t_0$. If so, we may pass to the limit under the sign of integration in (3) as $n \to \infty$, obtaining

(4) $$y = c + \int_0^t A(t_1)y \, dt_1$$

Differentiation yields $dy/dt = A(t)y$, and clearly $y(0) = c$.

We are deliberately violating our convention of representing the components of y by y_i because of our distaste for superscripts. There is at the moment no danger of confusion.

To demonstrate the convergence of the sequence $\{y_n\}$, we consider the series

(5) $$s(t) = \sum_{n=0}^{\infty} (y_{n+1} - y_n)$$

The Nth partial sum $s_N = \sum_{n=0}^{N} (y_{n+1} - y_n)$ has the simple form

$$s_N = y_{N+1} - y_0$$

Consequently, the series converges uniformly if and only if the sequence converges uniformly. The series will converge uniformly if the scalar series $\sum_{n=0}^{\infty} \|y_{n+1} - y_n\|$ converges uniformly. From the recurrence relation of (3) we obtain

(6) $$y_{n+1} - y_n = \int_0 A(t_1)(y_n - y_{n-1}) \, dt_1, \qquad n \geq 1$$

and thus

$$(7) \qquad \|y_{n+1} - y_n\| \le \int_0^t \|A(t_1)\| \, \|y_n - y_{n-1}\| \, dt_1, \qquad n \ge 1$$

Let $c_1 = \max \|A(t)\|$ for $0 \le t \le t_0$. Then (7) yields

$$(8) \qquad \|y_{n+1} - y_n\| \le c_1 \int_0^t \|y_n - y_{n-1}\| \, dt_1$$

Since $\|y_1 - y_0\| \le \int_0^t \|A(t_1)\| \, \|y_0\| \, dt_1 \le c_1 \|c\| t$, we obtain, inductively,

$$(9) \qquad \|y_{n+1} - y_n\| \le \|c\| \frac{(c_1 t)^{n+1}}{(n+1)!}, \qquad n = 0, 1, 2, \ldots$$

The uniform convergence of the exponential series in any finite interval ensures the uniform convergence of $\Sigma \|y_{n+1} - y_n\|$ and therefore that of the sequence $\{y_n\}$.

Notice that we make no attempt to prove that the sequence $\{dy_n/dt\}$ converges to dy/dt, but circumvent this difficulty by use of the integral equation of (4). This equation shows that the limit function is differentiable (which is not immediately obvious) and has the required derivative.

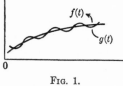

Fig. 1.

The use of integral equations to establish existence theorems is a standard device in the theory of differential equations, both ordinary and partial. It owes its efficiency to the smoothing properties of integration, as contrasted with coarsening properties of differentiation. If two functions are close (see Fig. 1), their integrals must be close, whereas their derivatives may be far apart and may not even exist.

Throughout the remaining chapters, we try wherever possible to convert the differential equations under consideration into integral equations. Very often, the key to the solution lies in the conversion to the proper integral equation.

Proof of Uniqueness. It is very important to prove uniqueness, since it is easy to construct equations which have multiple solutions. Naturally, in the latter case, $A(t)$ cannot be continuous.

Let z be another solution of (1), so that

$$(10) \qquad \frac{dz}{dt} = A(t)z, \qquad z(0) = c$$

for $0 \le t \le t_0$. Integration yields

$$(11) \qquad z = c + \int_0^t A(t_1)z \, dt_1$$

Combining this with (3), we obtain

$$(12) \qquad z - y_{n+1} = \int_0^t A(t_1)(z - y_n)\, dt_1$$

and hence

$$(13) \qquad \|z - y_{n+1}\| \leq \int_0^t \|A(t_1)\| \|z - y_n\|\, dt_1$$

Since $\|z - y_0\| \leq \|z\| + \|y_0\| \leq c_2 + \|c\|$, where $c_2 = \max \|z\|$ in $0 \leq t \leq t_0$, we obtain, via iteration,

$$(14) \qquad \|z - y_1\| \leq (c_2 + \|c\|)c_1 t$$

$$\cdot$$
$$\cdot$$
$$\cdot$$

$$\|z - y_{n+1}\| \leq (c_2 + \|c\|) \frac{(c_1 t)^{n+1}}{(n+1)!}$$

Letting $n \to \infty$, we obtain $\|z - y\| \leq 0$, whence $z \equiv y$.

Exercises

1. Show that the requirement that $A(t)$ be continuous may be replaced by the condition that $A(t)$ be Riemann-integrable.

2. What happens if $A(t)$ has an improper Riemann integral? Does there exist a solution? Is it unique? Consider $du/dt = u/2 \sqrt{t}$, $u(0) = 0$.

3. Show directly that the sequence $\{dy_n/dt\}$ converges uniformly, and then that the limit must be dy/dt.

---

4. The Matrix Equation, $dY/dt = A(t)Y$. Using precisely the same methods as above, we can prove that the matrix equation

$$(1) \qquad \frac{dZ}{dt} = A(t)Z, \qquad Z(0) = C$$

has a unique solution for $0 \leq t \leq t_0$. The details are left as an exercise.

In what follows, Y will be used to represent the solution of

$$(2) \qquad \frac{dY}{dt} = A(t)Y, \qquad Y(0) = I$$

where I is the identity matrix.

Exercise

1. Prove that $Z = YC$.

We shall occasionally make use of the following result:

Theorem 2. $Y(t)$ *is not singular in the interval* $[0,t_0]$. *More precisely,*

$$(3) \qquad |Y| = \exp \left\{ \int_0^t \Big[\sum_{i=1}^n a_{ii}(t_1) \Big] dt_1 \right\}$$

[The quantity $\sum_{i=1}^n a_{ii}$ occurs frequently in matrix theory and is therefore dignified by a special name, *trace*, written tr (A).]

Proof.   The proof depends upon the following two facts:

(4)   (a) $d|Y|/dt$ = sum of the determinants formed by replacing the elements of one row of $|Y|$ by their derivatives
     (b) The columns of Y are solutions of the vector equation

$$\frac{dy}{dt} = A(t)y$$

Simplifying the determinants obtained in (a) by use of (b), we obtain

$$(5) \qquad \frac{d}{dt} |Y| = \left(\sum_{i=1}^n a_{ii}(t) \right) |Y|$$

Since $|Y(0)| = 1$, (3) follows.

The fact that Y is nonsingular will play an important role in the solution of the inhomogeneous equation $dy/dt = A(t)y + w$.

Exercises

2. Prove that the solution of $dy/dt = A(t)y$, $y(0) = c$, is $y = Yc$.

3. (Alternative proof of the nonsingularity of Y.) Let y_i denote the ith column of Y. If $|Y| = 0$ at $t = t_1$, there exist nontrivial scalar constants $c_1, c_2, \ldots, c_n$, such that $c_1 y_1 + c_2 y_2 + \cdots + c_n y_n = 0$, the null vector, at $t = t_1$. Using the uniqueness theorem, show that this implies that $c_1 y_1 + c_2 y_2 + \cdots + c_n y_n = 0$ for $0 \le t \le t_0$, and that this is a contradiction at $t = 0$.

4. Let $u_1, u_2, \ldots, u_n$ be a set of n solutions of the nth order linear differential equation $u^{(n)} + a_1(t)u^{(n-1)} + \cdots + a_n(t)u = 0$. Show that the *Wronskian*

$$(6) \qquad w(t) = \begin{vmatrix} u_1 & u_2 & \cdots & u_n \\ u_1' & u_2' & \cdots & u_n' \\ \cdot & \cdot & & \cdot \\ \cdot & \cdot & & \cdot \\ \cdot & \cdot & & \cdot \\ u_1^{(n-1)} & u_2^{(n-1)} & \cdots & u_n^{(n-1)} \end{vmatrix}$$

is equal to $w(0) \exp \left[- \int_0^t a_1(t_1) \, dt_1 \right]$.

5. Prove that, if the coefficients are continuous in $[0,t_0]$, the linear differential equation in the above exercise has n solutions $u_1, u_2, \ldots, u_n$ with nonvanishing Wronskian in $[0,t_0]$. Hence show that no relation of the type $c_1u_1 + c_2u_2 + \cdots + c_nu_n = 0$, where the c_i are constants, can hold in $[0,t_0]$, and that any other solution in this interval may be expressed in the form $u = c_1u_1 + c_2u_2 + \cdots + c_nu_n$, where the c_i are constants.

6. Y^{-1} satisfies the equation $dZ/dt = -ZA(t)$. This is called the *adjoint equation*.

---

5. The Linear Inhomogeneous Equation $dy/dt = A(t)y + w$. Let us now consider the inhomogeneous equation

$$(1) \qquad \frac{dz}{dt} = A(t)z + w(t), \qquad z(0) = c$$

Let y denote the solution of the corresponding homogeneous equation,

$$(2) \qquad \frac{dy}{dt} = A(t)y, \qquad y(0) = c$$

and Y, as above, the solution of the matrix equation

$$(3) \qquad \frac{dY}{dt} = A(t)Y, \qquad Y(0) = I$$

To solve (1), we employ a method due to Lagrange, a variation of parameters. Let $z = Yu$. Substituting in (1),

$$(4) \qquad \frac{dz}{dt} = Y(t)\frac{du}{dt} + \frac{dY}{dt}u = Y(t)\frac{du}{dt} + A(t)Y(t)u$$
$$= A(t)Y(t)u + w(t)$$

Hence

$$(5) \qquad Y(t)\frac{du}{dt} = w$$

whence

$$(6) \qquad u = c + \int_0^t Y^{-1}(t_1)w(t_1)\, dt_1$$

[since $c = z(0) = Y(0)u(0) = u(0)$]. This yields for z the formula

$$(7) \qquad z = Y(t)c + \int_0^t Y(t)Y^{-1}(t_1)w(t_1)\, dt_1$$
$$= y + \int_0^t Y(t)Y^{-1}(t_1)w(t_1)\, dt_1$$

This result is important enough to distinguish as

Theorem 3. *The solution of* (1) *is given by*

$$(8) \qquad z = y + \int_0^t Y(t) Y^{-1}(t_1) w(t_1) \, dt_1$$

Exercises

1. Explain why our proof of Theorem 3 automatically implies the uniqueness of the solution of (1).

2. Use the Lagrange variation-of-parameters method to solve the inhomogeneous nth-order differential equation

$$u^{(n)} + a_1(t) u^{(n-1)} + \cdots + a_n(t) u = f(t)$$

(a) By converting it into an inhomogeneous system

(b) Directly by setting $u = \sum_{k=1}^{n} a_k(t) u_k(t)$, where $\{u_k(t)\}$ is a set of solutions of the homogeneous equation with nonvanishing Wronskian, and the $\{a_k(t)\}$ are unknowns

3. Prove directly that z as given by (8) is a solution of (1).

---

6. The Equation with Constant Coefficients. We now turn to the extremely important case where A is a constant matrix,

$$(1) \qquad \frac{dy}{dt} = Ay, \qquad y(0) = c$$

We shall show that this equation may be solved explicitly in terms of exponentials and polynomials. Before turning to this, let us derive some general properties of the solution of the matrix equation

$$(2) \qquad \frac{dY}{dt} = AY, \qquad Y(0) = I$$

Since $y = Yc$, it is Y which plays the important role. This representation shows very clearly the influence of initial condition upon the solution.

By analogy with the scalar equation $u' = au$, we are tempted to envisage a solution of (2) of the form $Y = e^{At}$. To give this formalism a meaning, we introduce the matrix series

$$(3) \qquad e^{At} = \sum_{n=0}^{\infty} \frac{A^n t^n}{n!}$$

where $A^0 = I$, which defines the matrix function on the left. Since $\|A^n\| \leq \|A\|^n$, $n = 1, 2, \ldots$, the m^2 series occurring on the right, where m is the order of A, are each majorized by $\sum_{n=0}^{\infty} \|A\|^n t^n / n!$. Hence these series converge uniformly in any finite interval, and the sum function is a continuous function of t for all finite t. Since the differentiated series converges uniformly, we have

$$(4) \qquad \frac{d}{dt} e^{At} = \sum_{n=1}^{\infty} \frac{A^n t^{n-1}}{(n-1)!} = A e^{At}$$

That $e^{At} = I$ for $t = 0$ is clear. Invoking the uniqueness theorem, we have that $Y(t) \equiv e^{At}$. In exactly the same manner as we derive from the series for e^t the scalar functional equation $e^{s+t} = e^s e^t$, we derive from the series representation for e^{At} that

$$(5) \qquad e^{At} e^{As} = \left(\sum_{n=0}^{\infty} \frac{A^n t^n}{n!} \right) \left(\sum_{n=0}^{\infty} \frac{A^n s^n}{n!} \right)$$

$$= \sum_{n=0}^{\infty} A^n \left(\sum_{k=0}^{n} \frac{t^k s^{n-k}}{k!(n-k)!} \right)$$

$$= \sum_{n=0}^{\infty} A^n \frac{(s+t)^n}{n!} = e^{A(s+t)}$$

The necessary rearrangements may be justified by the absolute convergence of the multiple series.

The following proof of the above functional equation for e^{At}, which depends upon the uniqueness theorem, is much more illuminating. Consider the two matrices $Y(t)Y(s)$ and $Y(s+t)$, where s is fixed and t is variable. Both satisfy the differential equation

$$(6) \qquad \frac{dZ}{dt} = AZ, \qquad Z(0) = Y(s)$$

since A is a constant matrix. Uniqueness requires that

$$(7) \qquad Y(t+s) = Y(t)Y(s)$$

Exercises

1. Prove that, if $Y(t+s) = Y(t)Y(s)$ for all s and t and if $Y(t)$ is continuous in an interval, then $dY/dt = AY$, where A is a constant matrix. (Pólya.)

2. Prove that $e^{(A+B)t} = e^{At} e^{Bt}$ for all t if and only if $AB = BA$.

3. Prove that $|e^{At}| = e^{t\,\mathrm{tr}\,(A)}$, and hence that e^{At} is never singular.

4. Prove that $(e^{At})^{-1} = e^{-At}$.

The functional equation (7) permits us to simplify (8) of Sec. 5 in the following important way:

Theorem 4. *If A is constant, the solution of*

$$(8) \qquad \frac{dz}{dt} = Az + w, \qquad z(0) = c$$

is given by

$$(9) \qquad z = y + \int_0^t Y(t - t_1) w(t_1) \, dt_1$$

where y is the solution of $dy/dt = Ay$, $y(0) = c$, and Y is the solution of $dY/dt = AY$, $Y(0) = I$.

7. The Behavior of the Solutions of $dy/dt = Ay$ as $t \to \infty$. Now that we have established some general properties of $Y(t)$, we shall focus our attention upon the individual elements of Y in order to determine the behavior of Y as $t \to \infty$. Let us note once and for all that when we write $t \to \infty$ we mean $t \to +\infty$.

There are several ways of determining the behavior of the individual terms. We shall begin with one fundamental procedure and then discuss the others subsequently.

Imitating the technique used for the nth-order linear differential equation, we set $y = e^{\lambda t} c$, where λ is a scalar constant and c is a constant vector which depends upon λ and which we desire to be nontrivial, that is, not equal to 0, the null vector. Substituting, we obtain

$$(1) \qquad \lambda e^{\lambda t} c = A e^{\lambda t} c$$

or

$$(2) \qquad \lambda c = Ac$$

This vector equation is equivalent to n linear homogeneous equations

$$(3) \qquad \sum_{j=1}^{n} a_{ij} c_j = \lambda c_i, \qquad i = 1, 2, \ldots, n$$

The necessary and sufficient condition that these equations possess a nontrivial solution for the c_i is the determinantal equation

$$(4) \qquad f(\lambda) = \begin{vmatrix} a_{11} - \lambda & a_{12} & \cdots & a_{1n} \\ a_{21} & a_{22} - \lambda & \cdots & a_{2n} \\ \cdot & \cdot & & \cdot \\ \cdot & \cdot & & \cdot \\ \cdot & \cdot & & \cdot \\ a_{n1} & a_{n2} & \cdots & a_{nn} - \lambda \end{vmatrix} = 0$$

This may be written simply as

$$(5) \qquad |A - \lambda I| = 0$$

This equation is called the *characteristic equation* of A, and its n roots $\lambda_1, \lambda_2, \ldots, \lambda_n$, in general complex, are called the *characteristic roots* of A.

Exercises

1. Prove that A and $T^{-1}AT$ have the same characteristic equation.
2. Prove that tr (A) is the sum of the characteristic roots of A.

For the next few sections we make the simplifying assumption that the characteristic roots of A are distinct.

This assumption permits us to establish the principal results in a very simple and elegant fashion. Subsequently we shall show how we may force most of our applications to depend only upon this simple case. In a later section we shall discuss the results corresponding to multiple roots.

To find the components $c_1, c_2, \ldots, c_n$ connected with a specific characteristic root λ_i, we take the cofactors of the element in any row of $|A - \lambda_i I|$. For example, taking the cofactors of the elements in the first row, we would have as possible choices for $c_1, c_2, \ldots, c_n$,

$$(6) \qquad c_1 = \begin{vmatrix} a_{22} - \lambda_i & a_{23} & \cdots & a_{2n} \\ a_{32} & a_{33} - \lambda_i & \cdots & a_{3n} \\ \cdot & & & \cdot \\ \cdot & & & \cdot \\ \cdot & & & \cdot \\ a_{n2} & a_{n3} & \cdots & a_{nn} - \lambda_i \end{vmatrix}$$

$$c_2 = - \begin{vmatrix} a_{21} & a_{23} & \cdots & a_{2n} \\ a_{31} & a_{33} - \lambda_i & \cdots & a_{3n} \\ \cdot & \cdot & & \cdot \\ \cdot & & & \\ \cdot & & & \\ a_{n1} & a_{n2} & \cdots & a_{nn} - \lambda_i \end{vmatrix}$$

$$c_n = \pm \begin{vmatrix} a_{21} & a_{22} - \lambda_i & \cdots & a_{2n-1} \\ a_{31} & a_{32} & \cdots & a_{3n-1} \\ \cdot & \cdot & & \cdot \\ \cdot & & & \\ \cdot & & & \cdot \\ a_{n1} & a_{n2} & \cdots & a_{nn-1} \end{vmatrix}$$

If, by mishap, all these cofactors vanish, we must try another set of cofactors. All tries cannot fail, since at least one of the cofactors of the elements, $a_{ii} - \lambda$, along the main diagonal must be distinct from zero. This is easily seen using the rule for differentiating a determinant. We have

(7) $\quad f'(\lambda) = -[\text{cofactor of } (a_{11} - \lambda)] - [\text{cofactor of } (a_{22} - \lambda)] - \cdots$
$$- [\text{cofactor of } (a_{nn} - \lambda)]$$

If all these cofactors were zero for some λ, then λ would be a root of $f'(\lambda) = 0$. which would imply that λ was a multiple root, in contradiction to our simplifying assumption above.

Let us note for future reference that the components $c_1, c_2, \ldots, c_n$ of c, which we shall call a *characteristic vector* of A, can always be chosen to be polynomials in the characteristic roots, as above.

Let $c^{(i)}$ be a characteristic vector associated with λ_i. We shall show in a moment that there can only be one such, apart from scalar multiples. To each characteristic root λ_i, there corresponds at least one solution, $y_i = e^{\lambda_i t} c^{(i)}$, of our differential equation $dy/dt = Ay$. Let us now demonstrate that these n solutions, corresponding to n distinct values of λ_i are linearly independent over any interval. Assume that there exists a relation

(8) $\qquad\qquad a_1 y_1 + a_2 y_2 + \cdots + a_n y_n = 0$

where $a_1, a_2, \ldots, a_n$ are scalar constants and 0 is the null vector. Differentiating k times, we obtain

(9) $\quad a_1 \lambda_1^k e^{\lambda_1 t} c^{(1)} + \cdots + a_n \lambda_n^k e^{\lambda_n t} c^{(n)} = 0, \qquad k = 0, 1, 2, \ldots, n-1$

Considering only the first components of the $c^{(i)}$, we obtain the equations

(10) $\qquad\qquad \sum_{i=1}^{n} a_i \lambda_i^k e^{\lambda_i t} c_1^{(i)} = 0, \qquad k = 0, 1, 2, \ldots, n-1$

In order that there may exist a nontrivial solution for the n quantities $a_i c_1^{(i)}$, it is necessary that the determinant

(11) $\qquad \begin{vmatrix} e^{\lambda_1 t} & e^{\lambda_2 t} & \cdots & e^{\lambda_n t} \\ \lambda_1 e^{\lambda_1 t} & \lambda_2 e^{\lambda_2 t} & \cdots & \lambda_n e^{\lambda_n t} \\ \cdot & \cdot & & \cdot \\ \cdot & \cdot & & \cdot \\ \cdot & \cdot & & \cdot \\ \lambda_1^{n-1} e^{\lambda_1 t} & \lambda_2^{n-1} e^{\lambda_2 t} & \cdots & \lambda_n^{n-1} e^{\lambda_n t} \end{vmatrix}$

be equal to zero. This, however, is impossible since the determinant is $[\exp (\lambda_1 + \lambda_2 + \cdots + \lambda_n)t]v(\lambda)$, where $v(\lambda)$ is the Vandermonde determinant of $\lambda_1, \lambda_2, \ldots, \lambda_n$, which is nonzero under our assumption concerning the distinctness of the λ_i. Therefore we must have $a_i c_1^{(i)} = 0$ for each i. If $a_i \neq 0$, we have $c_1^{(i)} = 0$. Continuing in this way, considering the other components of the $c^{(i)}$, we see that if $a_i \neq 0$, then $c^{(i)}$ must be the null vector. This cannot be, since $c^{(i)}$ is a characteristic vector which by construction is nontrivial.

A solution of the matrix equation $dZ/dt = AZ$ may now be obtained by forming the matrix Z whose columns are the y_i. That Z is nonsingular is a consequence of the linear independence demonstrated above. This is not immediately seen but proceeds from the following reasoning, which we have used before: If $|Z| = 0$ at $t = t_1$, we must have a relation of the form $a_1 y_1 + a_2 y_2 + \cdots + a_n y_n = 0$ at $t = t_1$, where 0 is again the null vector and the quantities $a_1, a_2, \ldots, a_n$ are scalars not all equal to zero. Since $y_1, y_2, \ldots, y_n$ are solutions of $dy/dt = Ay$, the linear combination $a_1 y_1 + a_2 y_2 + \cdots + a_n y_n$ is also a solution. It equals 0 at $t = t_1$, and consequently, by virtue of the uniqueness theorem, it must be identically equal to 0. From what has preceded, we see that actually each a_i is equal to zero, which is a contradiction. Thus $|Z| \neq 0$.

Since $Z(t)Z^{-1}(0)$ is a solution of $dY/dt = AY$, $Y(0) = I$, we must have $Y(t) = Z(t)Z^{-1}(0)$. Notice how useful the uniqueness theorem is for establishing identities between solutions.

This last relation shows very clearly the structure of $Y(t)$ and hence the structure of every solution of $dy/dt = Ay$, on the assumption of distinct characteristic roots. If A has multiple characteristic roots, the general solution may be more complicated, perhaps with polynomials in t appearing as coefficients.

Exercises

3. On the basis of the assumption that A has distinct characteristic roots, derive the necessary and sufficient condition that all the solutions of $dy/dt = Ay$ approach zero as $t \to \infty$.

4. Solve $t\, dy/dt = Ay$, where A is constant.

---

8. An Alternative Approach. In the preceding section we showed that the solutions y_i were linearly independent. Let us now show that the characteristic vectors $c^{(i)}$ are linearly independent. This is an immediate consequence of the previous independence result, since a relation of the type

$$(1) \qquad \sum_{i=1}^{n} a_i c^{(i)} = 0$$

is equivalent to $\displaystyle\sum_{i=1}^{n} a_i y_i = 0$ at $t = 0$.

This may also be proved by multiplying (1) by A repeatedly and using the fact that $Ac^{(i)} = \lambda_i c^{(i)}$. In this way we obtain the relations

$$(2) \qquad \sum_{i=1}^{n} a_i \lambda_i^k c^{(i)} = 0, \qquad k = 0, 1, 2, \ldots$$

and the argument now proceeds as above.

Exercises

1. Show that every solution of $dy/dt = Ay$ is a linear combination of the solutions $y_1, y_2, \ldots, y_n$.

2. Use this result to show that each λ_i possesses only one characteristic vector, apart from constant multiples, under our assumption of distinct characteristic roots.

Consider the matrix T formed by using the $c^{(i)}$ as columns. The linear independence of the $c^{(i)}$ is equivalent to $|T| \neq 0$. Since $Ac^{(i)} = \lambda_i c^{(i)}$, we have

$$(3) \qquad AT = T \begin{pmatrix} \lambda_1 & 0 & \cdots & 0 \\ 0 & \lambda_2 & \cdots & 0 \\ \cdot & \cdot & & \cdot \\ \cdot & \cdot & & \cdot \\ \cdot & \cdot & & \cdot \\ 0 & 0 & \cdots & \lambda_n \end{pmatrix}$$

whence

$$(4) \qquad T^{-1}AT = \begin{pmatrix} \lambda_1 & 0 & \cdots & 0 \\ 0 & \lambda_2 & \cdots & 0 \\ \cdot & \cdot & & \cdot \\ \cdot & \cdot & & \cdot \\ \cdot & \cdot & & \cdot \\ 0 & 0 & \cdots & \lambda_n \end{pmatrix}$$

A matrix of the type appearing on the right is called a *diagonal* matrix. This last result is important enough to state as a theorem:

Theorem 5. *If the characteristic roots $\lambda_1, \lambda_2, \ldots, \lambda_n$ of A are distinct, there exists a matrix T such that*

$$(5) \qquad T^{-1}AT = \begin{pmatrix} \lambda_1 & 0 & \cdots & 0 \\ 0 & \lambda_2 & \cdots & 0 \\ \cdot & \cdot & & \cdot \\ \cdot & \cdot & & \cdot \\ \cdot & \cdot & & \cdot \\ 0 & 0 & \cdots & \lambda_n \end{pmatrix}$$

Furthermore, the components of T can be chosen to be polynomials in the λ_i.

An application of this result is immediate. In the equation

(6) $$\frac{dY}{dt} = AY, \qquad Y(0) = I$$

set $Y = TZ$, with T as above. Then

(7) $$\frac{dZ}{dt} = T^{-1}ATZ, \qquad Z(0) = T^{-1}$$

or

(8) $$\frac{dZ}{dt} = \begin{pmatrix} \lambda_1 & 0 & \cdots & 0 \\ 0 & \lambda_2 & \cdots & 0 \\ \cdot & \cdot & & \cdot \\ \cdot & \cdot & & \cdot \\ \cdot & \cdot & & \cdot \\ 0 & 0 & \cdots & \lambda_n \end{pmatrix} Z, \qquad Z(0) = T^{-1}$$

The solution is clearly

(9) $$Z = \begin{pmatrix} e^{\lambda_1 t} & 0 & \cdots & 0 \\ 0 & e^{\lambda_2 t} & \cdots & 0 \\ \cdot & \cdot & & \cdot \\ \cdot & \cdot & & \cdot \\ 0 & 0 & \cdots & e^{\lambda_n t} \end{pmatrix} T^{-1}$$

whence

(10) $$Y = T \begin{pmatrix} e^{\lambda_1 t} & 0 & \cdots & 0 \\ 0 & e^{\lambda_2 t} & \cdots & 0 \\ \cdot & \cdot & & \cdot \\ \cdot & \cdot & & \cdot \\ 0 & 0 & \cdots & e^{\lambda_n t} \end{pmatrix} T^{-1}$$

Exercise

3. Derive (10) by using the relation $e^{At} = \sum_{n=0}^{\infty} A^n t^n / n!$.

---

9. The Jordan Canonical Form for Matrices with Multiple Characteristic Roots. In the previous section we showed how a matrix with distinct characteristic roots may be reduced to diagonal form. If A has multiple characteristic roots, it is not true, in general, that it may be transformed in this fashion. For example, it is easy to show that there exists no T such that

$$(1) \qquad T^{-1} \begin{pmatrix} 1 & 1 \\ 0 & 1 \end{pmatrix} T = \begin{pmatrix} 1 & 0 \\ 0 & 1 \end{pmatrix}$$

There is, however, an elegant canonical form for arbitrary matrices, with or without multiple characteristic roots, due to Jordan. Let

$$(2) \qquad L_k(\lambda) = \begin{pmatrix} \lambda & 1 & 0 & \cdots & 0 \\ 0 & \lambda & 1 & \cdots & 0 \\ \cdot & \cdot & \cdot & & \cdot \\ \cdot & \cdot & \cdot & & \cdot \\ \cdot & \cdot & \cdot & & 1 \\ 0 & 0 & 0 & \cdots & \lambda \end{pmatrix} \underbrace{}_{k}$$

If $k = 1$, we call this a simple factor.

The result of Jordan is that there exists a matrix T such that

$$(3) \qquad T^{-1} A T = \begin{pmatrix} L_{k_1}(\lambda_1) & 0 & \cdots & 0 \\ 0 & L_{k_2}(\lambda_2) & \cdots & 0 \\ \cdot & \cdot & & \cdot \\ \cdot & \cdot & & \cdot \\ \cdot & \cdot & & \cdot \\ 0 & 0 & \cdots & L_{k_r}(\lambda_r) \end{pmatrix}$$

where $k_1 + k_2 + \cdots + k_r = n$, and the λ_i are not necessarily distinct. For example, three possible types of 3×3 matrices with triple roots are

$$(4) \quad A_1 = \begin{pmatrix} \lambda_1 & 0 & 0 \\ 0 & \lambda_1 & 0 \\ 0 & 0 & \lambda_1 \end{pmatrix}, \quad A_2 = \begin{pmatrix} \lambda_1 & 0 & 0 \\ 0 & \lambda_1 & 1 \\ 0 & 0 & \lambda_1 \end{pmatrix}, \quad A_3 = \begin{pmatrix} \lambda_1 & 1 & 0 \\ 0 & \lambda_1 & 1 \\ 0 & 0 & \lambda_1 \end{pmatrix}$$

We shall not prove this classical result since, as mentioned above, we can reduce our problems to the case where A has distinct characteristic roots.

Exercises

1. Determine the form of e^{At} for general A.

2. Use this result to obtain the necessary and sufficient condition that $e^{At} \to 0$ as $t \to \infty$.

3. By considering the solution of $dy/dt = Ay$, where $A = \begin{pmatrix} 1 & 1 \\ 0 & 1 \end{pmatrix}$, show that A cannot be diagonalized.

4. Prove in the same way that A_1, A_2, and A_3 are distinct in the sense that there exists no T for which $T^{-1} A_i T = A_j$ for $i \neq j$.

5. Show that $L_k(\lambda) - \lambda I$, when raised to the kth power, yields the null matrix.

6. If A has distinct characteristic roots $\lambda_1, \lambda_2, \ldots, \lambda_n$, we may write n solutions of $dy/dt = Ay$ in the form $y = c(\lambda)e^{\lambda t}$, $\lambda = \lambda_1, \lambda_2, \ldots, \lambda_n$. If λ and μ are two distinct characteristic roots of A, then

$$\frac{c(\lambda)e^{\lambda t} - c(\mu)e^{\mu t}}{(\lambda - \mu)} = z$$

is also a solution of $dy/dt = Ay$. Is it true that, if λ is a multiple characteristic root, then the limit $\lim\limits_{\mu \to \lambda} z = \partial(c(\lambda)e^{\lambda t})/\partial\lambda = c'(\lambda)e^{\lambda t} + c(\lambda)te^{\lambda t}$ is also a solution of $dy/dt = Ay$? Generalize.

---

10. Another General Diagonalization Theorem. In the preceding section we discussed the Jordan canonical form of a matrix and its application to the representation of the general solution of the linear differential equation $dy/dt = Ay$. We have seen that a matrix may have a complicated canonical form if it possesses multiple characteristic roots. Let us now establish a result which is relatively easy to prove, which is useful, and which makes possible a semblance of order in this chaos.

Theorem 6. *There exists a matrix T having the property that*

$$(1) \qquad T^{-1}AT = \begin{pmatrix} \lambda_1 & b_{12} & \cdots & b_{1n} \\ 0 & \lambda_2 & \cdots & b_{2n} \\ \cdot & \cdot & & \cdot \\ \cdot & \cdot & & \cdot \\ \cdot & \cdot & & \cdot \\ 0 & 0 & \cdots & \lambda_n \end{pmatrix}$$

In the matrix on the right-hand side all the elements below the main diagonal are zero.

Proof. The proof proceeds by induction. Consider first 2×2 matrices. Let λ_1 be a characteristic root of A and $c^{(1)}$ be an associated characteristic vector. Let T be a matrix whose first column is $c^{(1)}$ and whose second column is chosen so that T is nonsingular. Then it follows that

$$(2) \qquad T^{-1}AT = \begin{pmatrix} \lambda_1 & b_{12} \\ 0 & b_{22} \end{pmatrix}$$

where we evaluate $T^{-1}AT$ most easily by taking it to be $T^{-1}(AT)$. Furthermore, b_{22} must equal λ_2, since $T^{-1}AT$ has the same characteristic roots as A.

Let us now demonstrate that the result for nth-order matrices may be used to establish the theorem for matrices of order $n + 1$. As before, let $c^{(1)}$ be a characteristic vector associated with λ_1, and let n other vectors $a^{(1)}$, $a^{(2)}$, . . . , $a^{(n)}$ be chosen so that the matrix T_1, whose columns are $c^{(1)}$, $a^{(1)}$, $a^{(2)}$, . . . , $a^{(n)}$, is nonsingular. As in the case of $n = 2$, we have

$$
(3) \quad T_1^{-1}AT_1 = \begin{pmatrix} \lambda_1 & b'_{12} & \cdots & b'_{1, n+1} \\ 0 & b'_{22} & \cdots & b'_{2, n+1} \\ \cdot & \cdot & & \cdot \\ \cdot & \cdot & & \cdot \\ \cdot & \cdot & & \cdot \\ 0 & b'_{n+1, 2} & \cdots & b'_{n+1, n+1} \end{pmatrix} = \begin{pmatrix} \lambda_1 & b'_{12} & \cdots & b'_{1, n+1} \\ 0 & & & \\ \cdot & & B_n & \\ \cdot & & & \\ 0 & & & \end{pmatrix}
$$

where B_n is an $n \times n$ matrix.

Since the characteristic equation of the right-hand side is

$$
(\lambda_1 - \lambda)|B_n - \lambda I| = 0
$$

it follows that the characteristic roots of B_n are λ_2, λ_3, . . . , λ_{n+1}, the remaining n characteristic roots of A. We know via the inductive hypothesis that there exists a nonsingular T_n such that

$$
(4) \quad T_n^{-1}B_nT_n = \begin{pmatrix} \lambda_2 & c_{12} & \cdots & c_{1n} \\ 0 & \lambda_3 & \cdots & c_{2n} \\ \cdot & \cdot & & \cdot \\ \cdot & \cdot & & \cdot \\ \cdot & \cdot & & \cdot \\ 0 & 0 & \cdots & \lambda_{n+1} \end{pmatrix}
$$

Let T_{n+1}, an $(n + 1) \times (n + 1)$ matrix, be formed as follows:

$$
(5) \quad T_{n+1} = \begin{pmatrix} 1 & 0 \\ 0 & T_n \end{pmatrix} = \begin{pmatrix} 1 & 0 & \cdots & 0 \\ 0 & & & \\ \cdot & & & \\ \cdot & & T_n & \\ \cdot & & & \\ 0 & & & \end{pmatrix}
$$

T_{n+1} is clearly nonsingular. We assert that

$$
(6) \quad T_{n+1}^{-1}(T_1^{-1}AT_1)T_{n+1} = \begin{pmatrix} \lambda_1 & b_{12} & \cdots & b_{1, n+1} \\ 0 & \lambda_2 & \cdots & b_{2, n+1} \\ \cdot & \cdot & & \cdot \\ \cdot & \cdot & & \cdot \\ \cdot & \cdot & & \cdot \\ 0 & 0 & \cdots & \lambda_{n+1} \end{pmatrix}
$$

Since $C^{-1}(B^{-1}AB)C = (BC)^{-1}A(BC)$, it follows that $T = T_1T_{n+1}$ is the required matrix of order $n + 1$. This completes the proof. Note again that the elements of T may be chosen to be polynomials in the characteristic roots.

Exercises

1. If A has k simple roots $\lambda_1, \lambda_2, \ldots, \lambda_k$, there exists a matrix T such that

$$T^{-1}AT = \begin{pmatrix} \lambda_1 & 0 & \cdots & 0 & b_{1,k+1} & \cdots & b_{1,n} \\ 0 & \lambda_2 & \cdots & 0 & b_{2,k+1} & \cdots & b_{2,n} \\ \cdot & \cdot & & \cdot & & & \cdot \\ \cdot & \cdot & & \cdot & & & \cdot \\ \cdot & \cdot & & \cdot & & & \cdot \\ 0 & 0 & \cdots & \lambda_k & b_{k,k+1} & \cdots & b_{k,n} \\ 0 & 0 & \cdots & 0 & \lambda_{k+1} & \cdots & b_{k+1,n} \\ \cdot & \cdot & & \cdot & & & \cdot \\ \cdot & \cdot & & \cdot & & & \cdot \\ \cdot & \cdot & & \cdot & & & \cdot \\ 0 & 0 & \cdots & 0 & 0 & \cdots & \lambda_n \end{pmatrix}$$

2. Is the T of Theorem 6 unique?

---

11. Corollary of the Above. We shall use below the following consequence of Theorem 6:

Corollary. *T may be chosen so that $\sum_{i,j} |b_{ij}|$ may be made less than any preassigned positive constant.*

At first sight this seems to contradict the result that not every matrix with multiple roots may be transformed into diagonal form. The point is that the T above depends upon the bound for $\Sigma|b_{ij}|$. If we attempt to choose a sequence of T's for which this bound goes to zero, we find that the sequence either approaches a singular matrix or has no limit.

Let T_1 be a matrix which reduces A to the semidiagonal form of Theorem 6. The change of variable $y = T_1z$ converts $dy/dt = Ay$ into

$$(1) \qquad \begin{aligned} \frac{dz_1}{dt} &= \lambda_1 z_1 + b_{12}z_2 + \cdots + b_{1n}z_n \\ \frac{dz_2}{dt} &= \qquad\quad \lambda_2 z_2 + \cdots + b_{2n}z_n \\ &\vdots \\ \frac{dz_n}{dt} &= \qquad\qquad\qquad\qquad \lambda_n z_n \end{aligned}$$

It is now easy to see how to choose T so as to satisfy the conditions of the corollary. Set $z_n = \epsilon^n z'_n$, $z_{n-1} = \epsilon^{n-1} z'_{n-1}$, $\ldots$, $z_1 = \epsilon z'_1$. The new system has the form

(2)
$$\frac{dz'_1}{dt} = \lambda_1 z'_1 + \epsilon b_{12} z'_2 + \cdots + \epsilon^{n-1} b_{1n} z'_n$$
$$\frac{dz'_2}{dt} = \qquad\quad \lambda_2 z_2 + \cdots + \epsilon^{n-2} b_{2n} z'_n$$
$$\vdots$$
$$\frac{dz'_n}{dt} = \qquad\qquad\qquad\qquad\quad \lambda_n z'_n$$

By suitable choice of ϵ the sum of the absolute values of the off-diagonal terms may be made as small as desired. This last transformation is equivalent to $z = Ez'$, where E is nonsingular. The required matrix is now $T = T_1 E$.

12. Application of the Previous Result. This last result may be used to obtain the form of the general solution of $dy/dt = Ay$, a result we had previously obtained on the basis of the unproved Jordan canonical form. Let T be a matrix which reduces A to semidiagonal form, and set $y = Tz$. The equation for z is

(1)
$$\frac{dz}{dt} = \begin{pmatrix} \lambda_1 & b_{12} & \cdots & b_{1n} \\ 0 & \lambda_2 & \cdots & b_{2n} \\ \cdot & \cdot & & \cdot \\ \cdot & \cdot & & \cdot \\ \cdot & \cdot & & \cdot \\ 0 & \cdot & \cdots & \lambda_n \end{pmatrix} z$$

Expressed in its components, this yields (1) of the previous section. Solving for the z_k one at a time, starting with z_n, we see that, if the λ_i are distinct, each z_i is a linear combination of exponentials. If, however, multiple characteristic roots appear, then these exponentials may have polynomials in t as coefficients. For example, if λ_{n-1} and λ_n are equal, the solution of

(2)
$$\frac{dz_{n-1}}{dt} = \lambda_{n-1} z_{n-1} + b_{n-1,\,n} z_n, \qquad \frac{dz_n}{dt} = \lambda_n z_n$$

yields terms of the form $z_n = c_1 e^{\lambda_n t}$, $z_{n-1} = (c_2 + c_3 t) e^{\lambda_n t}$, if $b_{n-1,\,n} \neq 0$. In general, we see that a characteristic root of multiplicity k gives rise to solutions with components of the form $P_{k-1}(t) e^{\lambda_k t}$, where $P_{k-1}(t)$ is a polynomial of degree $k - 1$ at most.

From this follows the important result:

Theorem 7. *The necessary and sufficient condition that all solutions of*

$$(3) \qquad \frac{dy}{dt} = Ay$$

tend to zero at $t \to \infty$ is that the real parts of the characteristic roots be negative.

Exercise

Show that the condition that the real parts of the characteristic roots be nonpositive is not sufficient to ensure that all the solutions of (3) be bounded as $t \to \infty$. What additional condition suffices?

---

13. An Approximation Theorem. The result which enables us to by-pass matrices with multiple characteristic roots is the following:

Theorem 8. *Given any matrix A, we can find a matrix B with distinct characteristic roots and such that $\|A - B\| \leq \epsilon$, where ϵ is any positive quantity.*

Proof. Consider the matrix $A + E$, where $E = (e_{ij})$ and the e_{ij} are independent real variables. If $A + E$ has a multiple characteristic root, then $f(\lambda) = |A + E - \lambda I|$ and $f'(\lambda)$ have a root in common. If $f(\lambda)$ and $f'(\lambda)$ have a root in common, their resultant $R(E)$ must vanish. We wish to show that we can find arbitrarily small values of e_{ij} for which $R(E) \neq 0$. If this is not true, $R(E)$ as a polynomial in the e_{ij} must vanish identically. This is impossible, since for the values $e_{ij} = -a_{ij}$, $i \neq j$, and $e_{ii} = i - a_{ii}$, $A + E$ does not have multiple roots. Hence $R(E) \neq 0$, identically.

We therefore can find e_{ij} for which $\displaystyle\sum_{i,j} |e_{ij}|$ is as small as desired and such that $A + E = B$ has distinct characteristic roots.

Exercises

1. Using this last result and the diagonalization theorem, find all continuous solutions of the functional equation $f(AB) = f(A)f(B)$, where f is a scalar function of A.

2. Prove Theorem 8 using the corollary in Sec. 11.

---

14. The Diagonalization of Variable Matrices. In Chap. 2, devoted to the asymptotic behavior of solutions of linear equations of the form

$$(1) \qquad \frac{dy}{dt} = (A + B(t))y$$

where A is a constant matrix and $B(t) \to 0$ as $t \to \infty$, it will be important to diagonalize $A + B(t)$. If the characteristic roots of A are distinct, it is plausible that, for t large enough, those of $A + B(t)$ will also be distinct.

In order to prove this result as well as some further results we shall require concerning the dependence of the characteristic roots of $A + B(t)$ upon t, we shall use a small amount of complex-variable theory. These results can certainly be proved without using this advanced analytic tool. However, the procedure we follow seems to be simultaneously the most natural, most elegant, and most informative, so that it would only be yielding to inverse snobbishness not to employ it.

Let $f(\lambda, t)$ be the characteristic polynomial of $A + B(t)$,

$$(2) \qquad f(\lambda, t) = |A + B(t) - \lambda I| = 0$$

and let $f(\lambda)$ be the characteristic polynomial of A, $f(\lambda) = f(\lambda, \infty)$. If the characteristic roots $\lambda_1, \lambda_2, \ldots, \lambda_n$ of A are distinct, we may, in the complex λ plane, draw circles c_i having the λ_i as centers, with the property that no two c_i have a point in common. The number of roots of $f(\lambda, t) = 0$ within c_i is given by

$$(3) \qquad N_i(t) = \frac{1}{2\pi i} \int_{c_i} \frac{\partial f(\lambda, t)/\partial \lambda}{f(\lambda, t)} \, d\lambda$$

As $t \to \infty$, $f(\lambda, t) = f(\lambda) + g(\lambda, t)$, where $g(\lambda, t) \to 0$, uniformly on any c_i. Hence for $t \geq t_1$, $f(\lambda, t)$ has no roots on a c_i. Since, also,

$$f_\lambda(\lambda, t) = f'(\lambda) + g_\lambda$$

we have

$$(4) \qquad N_i(t) = \frac{1}{2\pi i} \int_{c_i} \left[\frac{f'(\lambda) + g_\lambda}{f(\lambda) + g} \right] d\lambda$$

$$= \frac{1}{2\pi i} \int_{c_i} \frac{f'(\lambda)}{f(\lambda)} \, d\lambda + \frac{1}{2\pi i} \int_{c_i} \frac{f(\lambda)g_\lambda - f'(\lambda)g}{f(\lambda)f(\lambda, t)} \, d\lambda$$

Since $|f(\lambda)f(\lambda, t)|$ is uniformly bounded away from zero on c_i and since g and g_λ go to zero as t goes to infinity, we have, for $t \geq t_1$,

$$(5) \qquad N_i(t) = 1 + o(1)$$

Since $N_i(t)$ is an integer, it follows that $N_i(t) = 1$ for $t \geq t_1$. This establishes the distinctness of the characteristic roots of $A + B(t)$ for large t.

Let us now establish some further properties of the $\lambda_i(t)$, the characteristic roots of $A + B(t)$. We have, for $t \geq t_1$,

$$(6) \qquad \lambda_i(t) = \frac{1}{2\pi i} \int_{c_i} \frac{\lambda[\partial f(\lambda, t)/\partial \lambda]}{f(\lambda, t)} \, d\lambda$$

From the representation we see that, if $B(t)$ is continuous for $t \geq t_1$, then $\lambda_i(t)$ is continuous for $t \geq t_1$ and that $\lambda_i(t) \to \lambda_i$ as $t \to \infty$. If $B(t)$ is differentiable, the same must be true of the $\lambda_i(t)$.

Finally, if $B(t) \to 0$ as $t \to \infty$ and if $\int^{\infty} \|dB/dt\| \, dt < \infty$, then

$$\int^{\infty} \left| \frac{d\lambda_i(t)}{dt} \right| dt < \infty$$

This last follows readily from the fact that the coefficients of the various powers of λ in the polynomial $f(\lambda, t)$ are polynomials in the elements of $B(t)$.

For further reference, let us state these last results as

Theorem 9. *If $B(t) \to 0$ as $t \to \infty$ and if the characteristic roots of A are distinct, then, for $t \geq t_1$, the characteristic roots of $A + B(t)$ are distinct and approach those of A as $t \to \infty$. If, in addition, $\int^{\infty} \|dB/dt\| \, dt < \infty$, then $\int^{\infty} |d\lambda_i(t)/dt| \, dt < \infty$.*

Using this result, we can derive the following important tool:

Theorem 10. *Let*

(7) (a) *The characteristic roots of A be distinct*

  (b) $B(t) \to 0$ *as* $t \to \infty$

  (c) $\int^{\infty} \|dB/dt\| \, dt < \infty$

Then there exists a matrix $T(t)$ having the property that the change of variable $y = Tz$ converts

(8) $$\frac{dy}{dt} = (A + B(t))y$$

into

(9) $$\frac{dz}{dt} = (L(t) + C(t))z$$

where

(10) $$L(t) = \begin{pmatrix} \lambda_1(t) & 0 & \cdots & 0 \\ 0 & \lambda_2(t) & \cdots & 0 \\ \cdot & \cdot & & \cdot \\ \cdot & \cdot & & \cdot \\ \cdot & \cdot & & \cdot \\ 0 & 0 & \cdots & \lambda_n(t) \end{pmatrix}$$

and

(11) $$\int^{\infty} \|C(t)\| \, dt < \infty$$

Proof. $A + B(t)$, as noted above, has distinct characteristic roots for $t \geq t_1$. This fact permits us to find a matrix $T(t)$ which diagonalizes $A + B(t)$ for every sufficiently large t. Furthermore, we know that the elements of $T(t)$ can be chosen to be polynomials in the $\lambda_i(t)$. Since $\lambda_i(t) \to \lambda_i$ as $t \to \infty$, $T(t)$ can be chosen so that $T(t) \to T$, a nonsingular matrix. By virtue of our hypothesis concerning $B(t)$, every element t_{ij} of $T(t)$, has the property that $\int^{\infty} |dt_{ij}/dt|\, dt < \infty$. The substitution $y = Tz$ converts (8) into

$$(12) \qquad \frac{dz}{dt} = T^{-1}(t)[A + B(t)]T(t)z - T^{-1}(t)\frac{dT(t)}{dt}z$$

Since $T^{-1}(t)$ is uniformly bounded for $t \geq t_1$, we have

$$(13) \qquad \int^{\infty} \left\| T^{-1}(t)\frac{dT(t)}{dt} \right\| dt \leq c_1 \int^{\infty} \left\| \frac{dT(t)}{dt} \right\| dt < \infty$$

This completes the proof.

15. Linear Systems with Periodic Coefficients. In this section we consider systems of the form

$$(1) \qquad \frac{dy}{dt} = P(t)y$$

where $P(t)$ is a periodic matrix, which is to say that $P(t + \tau) = P(t)$, where τ is a nonzero, real constant.

Although these systems cannot be solved explicitly, as in the case where P is a constant matrix, we can find a representation for the general solution of (1) which is occasionally useful.

Theorem 11. *The solution of the matrix equation*

$$(2) \qquad \frac{dY}{dt} = P(t)Y, \qquad Y(0) = I,$$

where $P(t)$ is periodic with period τ and continuous for all t, has the form

$$(3) \qquad Y = Q(t)e^{Bt}$$

where B is a constant matrix and $Q(t)$ has period τ.

Proof. If Y has the stated form, then

$$(4) \qquad Y(t + \tau) = Q(t)e^{Bt}e^{B\tau} = Y(t)e^{B\tau}$$

Thus, it must be true that $Y^{-1}(t)Y(t + \tau) = e^{B\tau}$. Since $Y(t + \tau)$, as another solution of (2), is equal to $Y(t)C$, where C is a constant matrix, we must have $C = e^{B\tau}$. For B to exist, it is necessary that C be nonsingular.

This condition is satisfied, since $C = Y^{-1}(t)Y(t + \tau)$. We now show that this condition is sufficient.

Lemma. *If C is nonsingular, there exists a matrix B such that $e^B = C$.*

Proof of Lemma. If we use the Jordan normal form, the proof is quite simple. If C has the Jordan form

$$(5) \qquad C = T \begin{pmatrix} L_{k_1}(\lambda_1) & & 0 \\ & \cdot\ \cdot\ \cdot & \\ 0 & & L_{k_r}(\lambda_r) \end{pmatrix} T^{-1}$$

it is sufficient to show that each $L_k(\lambda)$ has a logarithm. For if B_i is a logarithm of $L_{k_1}(\lambda_i)$, then

$$(6) \qquad B = T \begin{pmatrix} B_1 & & 0 \\ & B_2 & \\ 0 & & B_r \end{pmatrix} T^{-1}$$

is a logarithm of C.

We utilize the result noted previously in Sec. 9, Exercise 5, that

$$(7) \qquad (L_k(\lambda) - \lambda I)^k = 0$$

the null matrix. From this it follows that the formal logarithm of $L_k(\lambda)$,

$$(8) \qquad \begin{aligned} B = \log L_k(\lambda) &= \log (\lambda I + L_k(\lambda) - \lambda I) \\ &= I \log \lambda + \sum_{n=1}^{\infty} \frac{(-1)^{n+1}}{n\lambda^n} (L_k(\lambda) - \lambda I)^n \\ &= I \log \lambda + \sum_{n=1}^{k-1} \frac{(-1)^{n+1}}{n\lambda^n} (L_k(\lambda) - \lambda I)^n \end{aligned}$$

exists and is actually a logarithm, as may be verified directly.

When B is determined so that $Y(t + \tau)Y^{-1}(t) = C = e^{B\tau}$, Q is determined by the relation $Q = Y(t)e^{-Bt}$. It is easily seen that Q is periodic of period τ.

Since we have not proved the validity of the Jordan normal form because of the long and tiresome argumentation required, let us present an independent proof of the above lemma.

We use first the result we have demonstrated above that any square matrix may be converted into triangular form; that is, there exists a T such that

$$(9) \qquad A = T^{-1} \begin{pmatrix} \lambda_1 & & \cdot\ \cdot\ \cdot \\ & \lambda_2 & \cdot\ \cdot \\ & & \cdot \\ 0 & & \lambda_N \end{pmatrix} T$$

where the notation signifies that all elements below the main diagonal are zero. If no multiple roots occur, we know that this triangular matrix may be taken to be diagonal, and it is clear that any nonsingular diagonal matrix has a logarithm. Consequently it is only the occurrence of multiple roots which provides any difficulty. Let us henceforth take A to be triangular.

The proof we present is inductive. The result is clearly true for 1×1 matrices. Let us assume that it holds for $n \times n$ matrices, with $n = 1, 2, \ldots, N$, and show that it holds for $(N + 1) \times (N + 1)$ matrices. Write A_{N+1} in the form

$$(10) \qquad A_{N+1} = \begin{pmatrix} A_N & a_N \\ 0 & \lambda_{N+1} \end{pmatrix}$$

where

$$(11) \qquad A_N = \begin{pmatrix} \lambda_1 & & \cdots \\ & \lambda_2 & \cdot\cdot \\ & & \cdot \\ 0 & & \lambda_N \end{pmatrix}$$

Here a_N is an N-dimensional column vector and 0 is an N-dimensional row vector with all zero elements.

Let B_N be a logarithm of A_N, the existence of B_N being given by the inductive hypothesis, and let

$$(12) \qquad B_{N+1} = \begin{pmatrix} B_N & x \\ 0 & l \end{pmatrix}$$

where 0 is as before, $l = \log \lambda_{N+1}$, and x is an unknown N-dimensional column vector. All logarithms of scalars that appear will be principal values, and thus any equation of the form $e^{\log a} = e^{\log b}$ will imply

$$\log a = \log b$$

It remains to be shown that x may be determined so that $e^{B_{N+1}} = A_{N+1}$. It is not difficult to verify by induction that

$$(13) \quad B_{N+1}^k = \begin{pmatrix} B_N^k & (B_N^{k-1} + B_N^{k-2}l + \cdots + l^{k-1}I)x \\ 0 & l^k \end{pmatrix} \quad k = 1, 2, \ldots$$

Hence

$$(14) \qquad e^{B_{N+1}} = \begin{pmatrix} e^{B_N} & \sum_{k=0}^{\infty} (B_N^{k-1} + B_N^{k-2}l + \cdots + l^{k-1}I)x/k! \\ 0 & \lambda_{N+1} \end{pmatrix}$$

The first two terms in the above sum are taken to be 1 and x. If l is not a characteristic root of B_N, we have

$$(15) \quad C(l) = \sum_{k=0}^{\infty} \frac{B_N^{k-1} + B_N^{k-2}l + \cdots + l^{k-1}I}{k!}$$

$$= \sum_{k=0}^{\infty} \frac{(B_N^k - l^k I)(B_N - lI)^{-1}}{k!} = (e^{B_N} - e^l I)(B_N - lI)^{-1}$$

Hence

$$(16) \quad |C(l)| = \frac{|e^{B_N} - e^l I|}{|B_N - lI|}$$

$$= \prod_{k=1}^{N} \frac{e^{r_k} - e^l}{r_k - l}$$

where $r_1, r_2, \ldots, r_N$ are the characteristic roots of B_N. From the definition of $C(l)$ we see that $|C(l)|$ is a continuous function of l. The right-hand side of (16) is also entire if we define $(e^{r_k} - e^l)/(r_k - l)$ in the obvious fashion at $l = r_k$. Consequently (16), which was proved under the hypothesis that $l \neq r_k$, holds for all l. It follows that $C(l)$ is never singular. Consequently we may determine x so that $C(l)x = a_N$, namely, $x = C(l)^{-1}a_N$.

Finally we observe that B_N will be real if the characteristic roots of A_N are positive.

Exercise

Find a representation for the solution of $u'' + p(t)u = 0$, where $p(t + 2\pi) = p(t)$.

BIBLIOGRAPHY

For the general theory of matrices we refer to:

Wedderburn, J. H. M., *Lectures on matrices*, Amer. Math. Soc. Colloquium Publications, vol. 17 (1934) (Dover reprint).

For the applications of matrix theory to differential equations,

Bellman, R., *A Survey of the theory of the boundedness, stability, and asymptotic behavior of solutions of linear and non-linear differential and difference equations*, Office of Naval Research, Washington, D.C., 1949.

Lefschetz, S., *Lectures on differential equations* (Annals of Mathematics Studies, no. 14), Princeton University Press, Princeton, N.J., 1946.

The contents of Sec. 14 are taken from:

Cesari, L., *Un nuovo criterio di stabilita per le soluzioni delle equazioni differenziali lineari*, Annali R. Scuola Norm. Sup. Pisa, ser. 2, vol. 9 (1940), pp. 163–186.

CHAPTER 2

STABILITY, BOUNDEDNESS, AND ASYMPTOTIC BEHAVIOR OF SOLUTIONS OF LINEAR SYSTEMS

1. Introduction. In this chapter we propose to consider the behavior of the solutions of the differential equation

$$(1) \qquad \frac{dz}{dt} = (A + B(t))z$$

where A is a constant matrix and $B(t)$ is small, in some sense, as $t \to \infty$. Two particularly important cases are those where $\|B(t)\| \to 0$ or where $\int^{\infty} \|B(t)\| \, dt < \infty$. Included in (1) are nth-order linear equations of the type

$$(2) \qquad \frac{d^n u}{dt^n} + (a_1 + p_1(t)) \frac{d^{n-1} u}{dt^{n-1}} + \cdots + (a_n + p_n(t))u = 0$$

and, in particular, the second-order equation

$$(3) \qquad \frac{d^2 u}{dt^2} + (a + p(t))u = 0$$

Because of its special form, many more results can be obtained concerning the solutions of (3) than for the solutions of (2) or (1). For that reason, we reserve a later chapter, Chap. 6, for a more complete discussion of (3), and in this chapter consider only those properties which are common to systems of all orders.

Intuitively, it seems reasonable to expect that the solutions of (1) should share many properties with the solutions of

$$(4) \qquad \frac{dy}{dt} = Ay$$

so far as their behavior as $t \to \infty$ is concerned. This expectation will be borne out to a great extent by many of the results we shall derive below. However, we shall also show, by means of counterexamples, that the behavior of the solutions of (1) is a good deal more complicated than one might suppose.

The problem arises as to the meaning of the word "small" as applied to $B(t)$ as $t \to \infty$. A first, and obvious, definition is that $B(t)$ is small if $\|B(t)\| \to 0$ as $t \to \infty$. This condition yields many interesting results. However, occasionally we shall require the stronger conditions

$$(5) \qquad\qquad \int^{\infty} \|B(t)\|\, dt < \infty$$

or

$$(5') \qquad\qquad \int^{\infty} \left\| \frac{dB(t)}{dt} \right\| dt < \infty$$

Essentially we see that the problem is that of imposing a suitable metric upon the space of variable matrices. Once this viewpoint has been taken, it becomes clear that the same procedure is advantageous in discussing the behavior of the solutions of (1) as $t \to \infty$. We shall investigate, variously,

$$(6) \qquad\qquad \varlimsup_{t \to \infty} \|z\|$$

or

$$(6') \qquad\qquad \lim_{t \to \infty} \|z\|$$

or, if these are infinite,

$$(6'') \qquad\qquad \lim_{t \to \infty} \frac{\log \|z\|}{t}$$

or, occasionally,

$$(6''') \qquad\qquad \int^{\infty} \|z\|^2\, dt$$

Each of these functionals has an important role to play in the study of the properties of the solutions of (1) for large values of t.

The point of the preceding discussion is that, in comparing the solutions of (1) with solutions of (4), we must agree to fasten our attention upon the class of perturbing matrices $B(t)$ we are admitting and upon the property of the solution in which we are interested. We may expect that some properties will be preserved under one class of perturbations and not under another, and this is indeed the case.

These preliminary remarks lead to a rigorous concept of stability for linear equations:

Definition. *The solutions of*

$$(7) \qquad\qquad \frac{dy}{dt} = A(t)y$$

*are stable with respect to a property P and perturbations B(t) of type **T** if the solutions of*

(8) $$\frac{dz}{dt} = (A(t) + B(t))z$$

also possess property P. If this is not true, the solutions of (7) are said to be unstable with respect to property P under perturbations of type T.

To illustrate this concept, consider the two simple differential equations

(9) $$\frac{du}{dt} = -au, \qquad \frac{dv}{dt} = (-a + b(t))v$$

where $a > 0$ and where $b(t) \to 0$ as $t \to \infty$. Both solutions have the properties

(10) (a) $\lim_{t \to \infty} u = \lim_{t \to \infty} v = 0$, a finite quantity

(b) $\lim_{t \to \infty} \dfrac{\log u}{t} = \lim_{t \to \infty} \dfrac{\log v}{t} = -a$

If, however, $a = 0$ and $b(t) = 1/t$, (10b) is preserved, but (10a) is not, since v is unbounded although u is bounded. Consequently, there is stability with respect to the property of (10b), but instability with respect to the property of boundedness. If we replace $1/t$ by a function which is integrable over (t_0, ∞), then boundedness will be preserved.

Perhaps the most important property of the solutions is that of boundedness. If a solution is bounded, we are interested in knowing whether or not it approaches zero as $t \to \infty$ and, in general, in examining the possible set of values it assumes as $t \to \infty$. If the solution is unbounded, we may wish to examine the ratio $(\log \|y\|)/t$ and perhaps other measures of its unboundedness.

2. Almost-constant Coefficients. We shall call the coefficient matrix $A(t)$ of the differential equation $dz/dt = A(t)z$ *almost constant* if

$$\lim_{t \to \infty} A(t) = A$$

a constant matrix. Our first results will concern the boundedness of solutions of equations of this type. Throughout the chapter we shall assume that the matrices that appear satisfy, in addition to the properties explicitly stated, the properties assumed for the existence and uniqueness of solutions.

Theorem 1. *If all solutions of*

(1) $$\frac{dy}{dt} = Ay$$

*where A is a constant matrix, are bounded as $t \to \infty$, the same is true of
the solutions of*

$$(2) \qquad \frac{dz}{dt} = (A + B(t))z$$

provided that $\int^{\infty} \|B(t)\| \, dt < \infty$.

Proof. We write equation (2) in the form

$$(3) \qquad \frac{dz}{dt} = Az + B(t)z$$

Identifying $B(t)z$ as an inhomogeneous term, we see, applying Theorem 4
of Chap. 1, that every solution of (3) satisfies a linear integral equation

$$(4) \qquad z = y + \int_0^t Y(t - t_1)B(t_1)z(t_1) \, dt_1$$

where y is the solution of (1) for which $y(0) = z(0)$ and where Y is the
matrix solution of

$$(5) \qquad \frac{dY}{dt} = AY, \qquad Y(0) = I$$

We note that $y = Yy(0) = Yz(0)$. Let $c_1 = \max\,(\sup_{t \geq 0} \|y\|, \sup_{t \geq 0} \|Y\|)$.
Then from (4) we obtain

$$(6) \qquad \begin{aligned} \|z\| &\leq \|y\| + \int_0^t \|Y(t - t_1)\| \|B(t_1)\| \|z(t_1)\| \, dt_1 \\ &\leq c_1 + c_1 \int_0^t \|B(t_1)\| \|z(t_1)\| \, dt_1 \end{aligned}$$

We now require the following lemma, of such utility throughout the
remainder of the book that we call it the *fundamental lemma:*

Lemma 1. *If $u,v \geq 0$, if c_1 is a positive constant, and if*

$$(7) \qquad u \leq c_1 + \int_0^t uv \, dt_1$$

then

$$(8) \qquad u \leq c_1 \exp\left(\int_0^t v \, dt_1\right)$$

Proof. From (7) we obtain

$$(9) \qquad \frac{uv}{c_1 + \int_0^t uv \, dt_1} \leq v$$

Integrating both sides between 0 and t,

$$(10) \qquad \log \left(c_1 + \int_0^t uv \, dt_1 \right) - \log c_1 \le \int_0^t v \, dt_1$$

or

$$(11) \qquad u \le c_1 + \int_0^t uv \, dt_1 \le c_1 \exp \left(\int_0^t v \, dt_1 \right)$$

Exercise

1. Show that (8) is the best possible consequence of (7).

Applying (11) to (6), we obtain

$$(12) \qquad \|z\| \le c_1 \exp \left(c_1 \int_0^t \|B\| \, dt_1 \right) \le c_1 \exp \left(c_1 \int_0^\infty \|B\| \, dt_1 \right)$$

Since, by assumption, $\int^\infty \|B\| \, dt_1 < \infty$, we see that $\|z\|$ is bounded.

Using the same technique, we may demonstrate

Theorem 2. *If all solutions of (1) approach zero as $t \to \infty$, the same holds for the solutions of (2), provided that $\|B(t)\| \le c_1$ for $t \ge t_0$, where c_1 is a constant which depends upon A.*

Proof. We have, as before,

$$(13) \qquad z = y + \int_0^t Y(t - t_1) B(t_1) z(t_1) \, dt_1$$

From the explicit representation of the solutions of $dy/dt = Ay$, it follows that, if $\|Y\| \to 0$ as $t \to \infty$, there exists a positive constant a such that $\|y\| \le c_2 e^{-at}$ and $\|Y(t)\| \le c_2 e^{-at}$ for $t \ge 0$. Hence

$$(14) \qquad \|z\| \le c_2 e^{-at} + c_2 \int_0^t e^{-a(t-t_1)} \|B(t_1)\| \|z(t_1)\| \, dt_1$$

or

$$(15) \qquad \|z\| e^{at} \le c_2 + c_1 c_2 \int_0^t e^{at_1} \|z(t_1)\| \, dt_1$$

Applying the fundamental lemma, we have

$$(16) \qquad \|z\| e^{at} \le c_2 e^{c_1 c_2 t}$$

If $c_1 c_2 < a$, we may conclude that $\|z\| \to 0$ as $t \to \infty$. Since the constants c_2 and a depend upon A, specifically upon the characteristic roots of A, it is clear that c_1 depends upon A.

Exercises

2. Consider the inhomogeneous equation $dz/dt = Az + w$, where A is a constant matrix and $w = w(t) \to w_0$, a constant vector, as $t \to \infty$. Discuss the boundedness of the solutions under the following alternate hypotheses:

(a) All solutions of $dy/dt = Ay$ approach zero as $t \to \infty$.

(b) All solutions of $dy/dt = Ay$ are unbounded.

(c) A has k characteristic roots with negative real parts.

3. What conditions on $w(t)$ will ensure that all solutions of

$$\frac{d^2u}{dt^2} + u = w(t)$$

are bounded as $t \to \infty$? Is $w(t) \to w_0$ as $t \to \infty$ sufficient?

3. Almost-constant Coefficients (Continued). We may improve upon Theorem 1 as follows:

Theorem 3. *Consider the system*

(1) $$\frac{dz}{dt} = (A + B(t) + C(t))z$$

where

(2) (a) *A is a constant matrix all of whose characteristic roots have non-positive real parts, while those with zero real parts are simple*

(b) $B(t) \to 0$ *as* $t \to \infty$, $\int^{\infty} \|dB/dt\| \, dt < \infty$

(c) $\int^{\infty} \|C(t)\| \, dt < \infty$

(d) *The characteristic roots of $A + B(t)$ have nonpositive real parts for $t \geq t_0$*

Under these conditions, all solutions of (1) are bounded as $t \to \infty$.

Proof. Let the simple characteristic roots of A which possess zero real parts be denoted by $\lambda_1, \lambda_2, \ldots, \lambda_k$. Then, as indicated in Exercise 1 of Sec. 10, Chap. 1, there exists a matrix T such that

(3) $$T^{-1}AT = \begin{pmatrix} \lambda_1 & & & 0 & d_{1k+1} & \cdots & d_{1n} \\ & \lambda_2 & & & & & \\ & & \ddots & & & & \\ 0 & & & \lambda_k & d_{kk+1} & \cdots & d_{kn} \\ & & & & \lambda_{k+1} & & \\ & 0 & & & & \ddots & \\ & & & & & & \lambda_n \end{pmatrix}$$

The notation signifies that the initial $k \times k$ submatrix is diagonal and that all the elements beneath the main diagonal are zero. As usual, the elements of T are chosen to be polynomials in the elements and characteristic roots of A.

Let us turn our attention now to the matrix $A + B(t)$, whose characteristic roots we designate by $\lambda_1(t), \lambda_2(t), \ldots, \lambda_n(t)$. Let $\lambda_1(t)$,

$\lambda_2(t), \ldots, \lambda_k(t)$ be the characteristic roots of $A + B(t)$, which approach $\lambda_1, \lambda_2, \ldots, \lambda_k$ and which we know to be simple for t sufficiently large. The remaining characteristic roots $\lambda_{k+1}(t), \ldots, \lambda_n(t)$ have negative real parts for large t, and uniformly so as $t \to \infty$.

Let $T(t)$ be the matrix corresponding to T above, formed so that $T^{-1}(t)[A + B(t)]T(t)$ has a form similar to (3) with $\lambda_i(t) \to \lambda_i$. In this way we have $T(t) \to T$ as $t \to \infty$ and therefore are assured that $T^{-1}(t)$ is uniformly bounded as $t \to \infty$. Furthermore the assumption that $\int^{\infty} \|dB/dt\| \, dt < \infty$ yields the result that $\int^{\infty} \|dT/dt\| \, dt < \infty$. Both of these facts are important for what follows.

We now make the substitution $z = T(t)w$ in (1), obtaining

$$(4) \qquad \frac{dw}{dt} = T^{-1}(A + B(t))Tw + \left(T^{-1}CT - T^{-1}\frac{dT}{dt} \right)w$$

By virtue of our hypothesis concerning C and of the results above concerning T, we see that the matrix $T^{-1}CT - T^{-1}\,dT/dt$, which we call R, is absolutely integrable,

$$(5) \qquad \int^{\infty} \|R\| \, dt = \int^{\infty} \|T^{-1}CT - T^{-1}\,dT/dt\| \, dt < \infty$$

Writing equation (4) out in terms of its components, we have

$$(6) \quad (a) \quad \frac{dw_i}{dt} = \lambda_i(t)w_i + \sum_{j=k+1}^{n} d_{ij}(t)w_j + \sum_{j=1}^{n} r_{ij}(t)w_j, \, i = 1, 2, \ldots, k$$

$$(b) \quad \frac{dw_i}{dt} = \lambda_i(t)w_i + \sum_{j=i+1}^{n} d_{ij}(t)w_j + \sum_{j=1}^{n} r_{ij}(t)w_j, \, i = k + 1, \ldots, n$$

where $d_{ij}(t) \to d_{ij}$, a constant, as $t \to \infty$, and where $\int^{\infty} |r_{ij}(t)| \, dt < \infty$.

Note that the only difference between (6a) and (6b) resides in the summation over the terms $d_{ij}w_j$, a difference which is a consequence of the form of (3).

Let us now discuss the solution of (6b). Taking the case of $i = n$ first, we find that

$$(7) \quad w_n = c_n \exp \left[\int_0^t \lambda_n(t_1) \, dt_1 \right]$$

$$+ \int_0^t \exp \left[\int_{t_1}^t \lambda_n(s) \, ds \right] \left(\sum_{j=1}^{n} r_{nj}(t_1)w_j(t_1) \right) dt_1$$

Since the real parts of the $\lambda_i(t)$ for $i = k + 1, \ldots, n$ are uniformly nonnegative, we have, for some positive constant a,

$$(8) \qquad |w_n| \leq |c_n|e^{-at} + \int_0^t e^{-a(t-t_1)}\|R\|\,\|w\|\,dt_1$$

Considering the case $i = n - 1$, we obtain the integral equation

$$(9) \quad w_{n-1} = c_{n-1}\exp\left[\int_0^t \lambda_{n-1}(t_1)\,dt_1\right]$$
$$+ \int_0^t \exp\left[\int_{t_1}^t \lambda_{n-1}(s)\,ds\right]d_{nn}(t_1)w_n(t_1)\,dt_1$$
$$+ \int_0^t \exp\left[\int_{t_1}^t \lambda_n(s)\,ds\right]\left(\sum_{j=1}^n r_{n-1,\,j}(t_1)w_j(t_1)\right)dt_1$$

This yields the inequality

$$(10) \quad |w_{n-1}| \leq |c_{n-1}|e^{-at} + \int_0^t e^{-a(t-t_1)}|d_{nn}(t_1)|\,|w_n(t_1)|\,dt_1$$
$$+ \int_0^t e^{-a(t-t_1)}\|R\|\,\|w\|\,dt_1$$

If we employ the previous inequality, (8), we obtain

$$(11) \quad |w_{n-1}| \leq |c_{n-1}|e^{-at} + \int_0^t e^{-a(t-t_1)}\|R\|\,\|w\|\,dt_1$$
$$+ c_{n+1}\int_0^t e^{-a(t-t_1)}\left[|c_n|e^{-at_1} + \int_0^{t_1} e^{-a(t_1-t_2)}\|R\|\,\|w\|\,dt_2\right]dt_1$$

where we have used the fact that $|d_{nn}(t_1)| \leq c_{n+1}$. The first term in the second integral yields $c_{n+1}|c_n|te^{-at}$. The second term is

$$(12) \quad c_{n+1}\int_0^t e^{-a(t-t_1)}\left(\int_0^{t_1} e^{-a(t_1-t_2)}\|R\|\,\|w\|\,dt_2\right)dt_1$$
$$= c_{n+1}e^{-at}\int_0^t\left(\int_0^{t_1} e^{at_2}\|R\|\,\|w\|\,dt_2\right)dt_1$$

Integrated by parts, this becomes

$$(13) \qquad c_{n+1}\int_0^t (t - t_1)e^{-a(t-t_1)}\|R\|\,\|w\|\,dt_1$$

Since $te^{-at} \leq b_1 e^{-a_1 t}$ for $t \geq 0$ if $a_1 < a$ and if b_1 is suitably chosen, we obtain

$$(14) \qquad |w_{n-1}| \leq b_2 e^{-a_1 t} + b_3\int_0^t e^{-a_1(t-t_1)}\|R\|\,\|w\|\,dt_1$$

Since $a_1 < a$, we obtain the same inequality for $|w_n|$ by increasing the values of b_2 and b_3, if necessary.

Continuing in this way step by step, we find that constants b_4, b_5, and a_1 exist such that

$$(15) \qquad |w_i| \le b_4 e^{-a_1 t} + b_5 \int_0^t e^{-a_1(t-t_1)} \|R\| \|w\| \, dt_1$$

for $k + 1 \le i \le n$.

Turning to the equations where $1 \le i \le k$, we obtain

$$(16) \quad w_i = c_i \exp\left[\int_0^t \lambda_i(t_1) \, dt_1 \right]$$
$$+ \int_0^t \exp\left[\int_{t_1}^t \lambda_i(t_2) \, dt_2 \right] \left(\sum_{j=k+1}^n d_{ij}(t_1) w_j \right) dt_1$$
$$+ \int_0^t \exp\left[\int_{t_1}^t \lambda_i(t_2) \, dt_2 \right] \left(\sum_{j=1}^n r_{ij}(t_1) w_j \right) dt_1$$

Since the real parts of $\lambda_i(t)$ are nonpositive, we have, for $i = 1, 2, \ldots, k$,

$$(17) \qquad |w_i| \le |c_i| + c_3' \int_0^t \left(\sum_{j=k+1}^n |w_j| \right) dt_1 + \int_0^t \|R\| \|w\| \, dt_1$$

From (15) we have

$$(18) \qquad \sum_{j=k+1}^n |w_j| \le c_5' e^{-a_1 t} + c_6' \int_0^t e^{-a_1(t-t_1)} \|R\| \|w\| \, dt_1$$

Since

$$(19) \quad \int_0^t \left(\int_0^{t_1} e^{-a_1(t_1-t_2)} \|R\| \|w\| \, dt_2 \right) dt_1$$
$$= -\frac{1}{a_1} \int_0^t e^{-a_1(t-t_1)} \|R\| \|w\| \, dt_1 + \frac{1}{a_1} \int_0^t \|R\| \|w\| \, dt_1$$

we obtain, finally, from (17),

$$(20) \qquad |w_i| \le c_7' + c_8' \int_0^t \|R\| \|w\| \, dt_1, \qquad 1 \le i \le k$$

Combining this with (15), we derive

$$(21) \qquad \|w\| \le c_9' + c_{10}' \int_0^t \|R\| \|w\| \, dt_1$$

Since by assumption $\int^\infty \|R\| \, dt < \infty$, it follows by application of the fundamental lemma that $\|w\|$ is bounded.

Exercises

1. Making use of the Jordan canonical form, show that the hypothesis that the roots with zero real part are simple may be replaced by the condition that they correspond to simple factors.

2. For what values of c_1, c_2, and c_3 are all solutions of

$$\frac{d^2u}{dt^2} + \frac{c_1}{t}\frac{du}{dt} + \left(1 + \frac{c_2}{t} + \frac{c_3}{t^2}\right)u = 0$$

bounded?

---

4. Equations with Periodic Coefficients. We now consider equations of the form

$$(1) \qquad \frac{dz}{dt} = (A(t) + B(t))z$$

where $A(t)$ is periodic and $B(t)$ is small as $t \to \infty$. That the stability properties derived for the case where A is constant carry over to this case is a consequence of the canonical representation of solutions of the unperturbed equation

$$(2) \qquad \frac{dy}{dt} = A(t)y$$

furnished by Theorem 11 of Chap. 1, which tells us that the matrix solution of

$$(3) \qquad \frac{dY}{dt} = A(t)Y, \qquad Y(0) = I$$

has the form

$$(4) \qquad Y(t) = P(t)e^{ct}$$

where $P(t)$ has the same period as $A(t)$ and where C is a constant matrix.

Using this representation, we may prove quite readily

Theorem 4. *If all solutions of* (2) *are bounded, then all solutions of* (1) *are also bounded, provided that*

(5)   (a) $A(t)$ *is periodic*

       (b) $\displaystyle\int^{\infty} \|B\|\, dt < \infty$

If all solutions of (2) *approach zero, as* $t \to \infty$, *then the same holds for all solutions of* (1), *under the same hypotheses.*

Proof.   If all solutions of (2) are bounded, we must have $\|e^{ct}\|$ bounded as $t \to \infty$; if all solutions of (2) tend to zero, then $\|e^{ct}\| \to 0$ and does so exponentially, which is to say, $\|e^{ct}\| \le c_1 e^{-at}$, with $a > 0$.   Since

$$(6) \qquad z = y + \int_0^t Y(t) Y^{-1}(t_1) B(t_1) z(t_1)\, dt_1$$

$$= y + \int_0^t P(t) e^{c(t-t_1)} P(t_1)^{-1} B(t_1) z(t_1)\, dt_1$$

we have

$$(7) \qquad \|z\| \le \|y\| + \int_0^t \|P(t)\| \|e^{c(t-t_1)}\| \|P(t_1)^{-1}\| \|B(t_1)\| \|z(t_1)\|\, dt_1$$

$$\le c_1 + c_1 \int_0^t \|B(t_1)\| \|z(t_1)\|\, dt_1$$

whence boundedness follows, as before.   The second part of the statement is also derived along previous lines.

5. Equations with General Variable Coefficients.   We already know that the boundedness of the solutions of

$$(1) \qquad\qquad \frac{dy}{dt} = A(t) y$$

together with the condition $\|B(t)\| \to 0$ as $t \to \infty$ is not sufficient to ensure the boundedness of all solutions of

$$(2) \qquad\qquad \frac{dz}{dt} = (A(t) + B(t)) z$$

[compare (10) of Sec. 1].   We might be tempted, in the light of preceding results, to state that the result will be valid provided that we amend the condition $\|B(t)\| \to 0$ to read $\int^{\infty} \|B(t)\|\, dt < \infty$.   Let us show by a counterexample that no such general theorem can hold.

Theorem 5.   *There is an equation of type* (1) *with the property that all solutions approach zero as $t \to \infty$, and a matrix $B(t)$ for which* $\int^{\infty} \|B(t)\|\, dt < \infty$, *such that all solutions of* (2) *are not bounded.*

Proof.   Consider the equation

$$(3) \qquad\qquad \frac{dy_1}{dt} = -a y_1$$

$$\frac{dy_2}{dt} = (\sin \log t + \cos \log t - 2a) y_2$$

whose general solution is

$$(4) \qquad\qquad y_1 = c_1 e^{-at}$$

$$y_2 = c_2 e^{t \sin \log t - 2at}$$

If $a > \frac{1}{2}$, every solution approaches zero as $t \to \infty$. If we choose as our perturbing matrix

$$(5) \qquad B(t) = \begin{pmatrix} 0 & 0 \\ e^{-at} & 0 \end{pmatrix}$$

the perturbed equation has the form

$$(6) \qquad \frac{dz_1}{dt} = -az_1$$

$$\frac{dz_2}{dt} = (\sin \log t + \cos \log t - 2a)z_2 + z_1 e^{-at}$$

The solution of this system is

$$(7) \qquad z_1 = c_1 e^{-at}$$

$$z_2 = e^{t \sin \log t - 2at}\left(c_2 + c_1 \int_0^t e^{-t_1 \sin \log t_1}\, dt_1\right)$$

Let $t = e^{(2n+\frac{1}{2})\pi}$. Since

$$(8) \qquad \int_0^t e^{-t_1 \sin \log t_1}\, dt_1 > \int_{te^{-\pi}}^{te^{-2\pi/3}} e^{-t_1 \sin \log t_1}\, dt_1$$

$$> t(e^{-2\pi/3} - e^{-\pi}) \exp\left(-\frac{e^{-\pi}t}{2}\right)$$

we see that, if

$$(9) \qquad 1 < 2a < 1 + e^{-\pi/2}$$

the solutions of (6) will be bounded only if $c_1 = 0$. This condition is fulfilled only for those solutions for which $z_1(0) = 0$.

Let us now turn to the problem of seeing what we can salvage from this. We can prove

Theorem 6. *If all the solutions of (1) are bounded, then all the solutions of (2) are bounded, provided that*

$$(10) \quad (a)\ \int^\infty \|B(t)\|\, dt < \infty$$

$$(b)\ \lim_{t \to \infty} \int^t \mathrm{tr}\,(A)\, dt > -\infty$$

or, in particular, that

$$(b')\ \mathrm{tr}\,(A) = 0$$

Condition (b') is relevant to the important equation

$$(11) \qquad u'' + a(t)u = 0$$

which is equivalent to a two-dimensional system satisfying (b').

Proof. Expressing z in terms of y, we have

$$(12) \qquad z = y + \int_0^t Y(t) Y^{-1}(t_1) B(t_1) z(t_1) \, dt_1$$

whence

$$(13) \qquad \|z\| \leq \|y\| + \int_0^t \|Y(t)\| \, \|Y^{-1}(t_1)\| \, \|B(t_1)\| \, \|z(t_1)\| \, dt_1$$

Since

$$(14) \qquad \det Y = \exp \left[\int_0^t \operatorname{tr} (A) \, dt \right]$$

we see that, provided (10b) is satisfied, $\|Y^{-1}(t)\|$ is bounded as $t \to \infty$. Hence from (13), we obtain

$$(15) \qquad \|z\| \leq c_1 + c_1 \int_0^t \|B(t_1)\| \, \|z(t_1)\| \, dt_1$$

and application of the fundamental lemma yields the desired boundedness.

6. Almost-constant Coefficient : Asymptotic Behavior. We now return to equations of the form

$$(1) \qquad \frac{dz}{dt} = (A + B(t))z$$

where $\|B(t)\| \to 0$ as $t \to \infty$, and investigate the behavior of $\|z\|$ as $t \to \infty$. The simplest, and perhaps the most interesting, case is that where A has simple characteristic roots. For the case of multiple roots there are corresponding results which are more complicated to state and prove. Consequently, we content ourselves with the following:

Theorem 7. *If, in equation (1), the following conditions are satisfied:*

(2) *(a) A is a constant matrix with simple characteristic roots*
        (b) $\|B\| \to 0$ as $t \to \infty$

then, corresponding to any characteristic root λ_k, there is a solution $z^{(k)}$ satisfying the inequalities

$$(3) \quad c_2 \exp \left[\operatorname{Re} (\lambda_k)t - d_2 \int_{t_0}^t \|B\| \, dt \right] \leq \|z^{(k)}\|$$
$$\leq c_1 \exp \left[\operatorname{Re} (\lambda_k)t + d_1 \int_{t_0}^t \|B\| \, dt \right]$$

for $t \geq t_0$, with c_1, c_2, d_1, and d_2 being positive constants.
In particular,

$$(4) \qquad \lim_{t \to \infty} \frac{\log \|z^{(k)}\|}{t} = \operatorname{Re} (\lambda_k)$$

Furthermore, if the characteristic roots are real and distinct, and if $\int^{\infty} \|B\| \, dt < \infty$, *there are n solutions* $z^{(1)}, z^{(2)}, \ldots, z^{(n)}$, *such that*

$$(5) \qquad z^{(k)} = e^{\lambda_k t}(c_k + o(1))$$

as $t \to \infty$, *where* c_k *is a constant vector.*

Proof. Let C be a constant matrix reducing A to diagonal form, $C^{-1}AC = L$, where L is diagonal. The substitution $z \to Cz$ transforms (1) into an equation of the same form where A is now diagonal, and where the new $B(t) \to 0$ as $t \to \infty$. Hence we start with the equation in this form, since it is clear that, if the solutions of this new equation exist having the desired property, then the original equation also possesses solutions of the stated type.

A slight bit of complication is furnished by the fact that, even though the λ_k are distinct, the real parts may coincide. Let λ_k be a characteristic root satisfying $\cdots$ Re $(\lambda_{k-1}) \leq$ Re $(\lambda_k) <$ Re $(\lambda_{k+1}) \leq$ Re (λ_{k+2}) $\cdots$, and let y_k be the column vector whose components are $0, 0, \ldots,$ $e^{\lambda_k t}, 0, \ldots, 0$, the $e^{\lambda_k t}$ occurring in the kth position. Any solution of (1) satisfies the integral equation

$$(6) \qquad z = y + \int_0^t Y(t - t_1)B(t_1)z(t_1) \, dt_1$$

where, as usual, $Y(t) = \exp Lt$; and conversely any solution of (6) is a solution of the differential equation. To obtain a solution to (6) satisfying (4), we must somehow eliminate the terms involving $e^{\lambda_l t}$ for $l > k$ (which is, of course, not necessary if $\lambda_k = \lambda_n$). This we do as follows: Decompose Y into the sum $Y = Y_1 + Y_2$, where

$$(7) \qquad Y_1 = \begin{pmatrix} e^{\lambda_1 t} & & & & & & & 0 \\ & e^{\lambda_2 t} & & & & & & \\ & & \ddots & & & & & \\ & & & e^{\lambda_k t} & & & & \\ & & & & 0 & & & \\ & & & & & \ddots & & \\ 0 & & & & & & & 0 \end{pmatrix}$$

$$Y_2 = \begin{pmatrix} 0 & & & & & & 0 \\ & \ddots & & & & & \\ & & 0 & & & & \\ & & & e^{\lambda_{k+1} t} & & & \\ & & & & e^{\lambda_{k+2} t} & & \\ & & & & & \ddots & \\ 0 & & & & & & e^{\lambda_n t} \end{pmatrix}$$

Using this decomposition, (6) takes the form

$$(8) \quad z = y + \int_0^t Y_1(t - t_1)B(t_1)z(t_1)\, dt_1 + \int_0^t Y_2(t - t_1)B(t_1)z(t_1)\, dt_1$$

The undesirable terms are now united in the second integral. To eliminate these terms, we use the fact that $Y_2(t - t_1)B(t_1)z(t_1)$ is a solution of $dy/dt = Ay$, for any fixed t_1, and consequently, provided that the integral converges, so also is $\int_0^\infty Y_2(t - t_1)B(t_1)z(t_1)\, dt_1$. Hence, by changing our choice of a particular y, but still regarding it as a generic solution of $dy/dt = Ay$, we may write (8) in the form

$$(9) \quad z = y + \int_0^t Y_1(t - t_1)B(t_1)z(t_1)\, dt_1 - \int_t^\infty Y_2(t - t_1)B(t_1)z(t_1)\, dt_1$$

It is, of course, no longer immediate that a solution of (9) exists. To obtain a solution, we have recourse to the method of successive approximations. We shall show that, with a suitable choice of y, a solution of (9) exists satisfying the right-hand inequality of (3). To obtain the left inequality, another device is required.

Choose y to be the vector y_k defined above, and define

$$(10) \quad z_0 = y_k$$
$$z_{n+1} = y_k + \int_0^t Y_1(t - t_1)B(t_1)z_n(t_1)\, dt_1 - \int_t^\infty Y_2(t - t_1)B(t_1)z_n(t_1)\, dt_1$$

The point t_0 is chosen in place of zero, since we shall require subsequently that $\|B(t)\|$ be uniformly small for $t \geq t_0$.

Let us first show by induction that

$$(11) \quad \|z_n\| \leq c_1 \exp\left[\text{Re}\,(\lambda_k)t + d_1 \int_{t_0}^t \|B\|\, dt\right]$$

for a suitable $c_1 \geq 1$, $d_1 \geq 0$.

The result clearly holds for $n = 0$. Let us assume that it is valid for n, and show that this implies its truth for $n + 1$. We have

$$(12) \quad \|z_{n+1}\| \leq \|y_k\| + \int_{t_0}^t \|Y_1(t - t_1)\|\,\|B(t_1)\|\,\|z_n\|\, dt_1$$
$$+ \int_t^\infty \|Y_2(t - t_1)\|\,\|B(t_1)\|\,\|z_n\|\, dt_1$$

The first integral is bounded by

$$(13) \quad c_1 \int_{t_0}^t \|Y_1(t - t_1)\|\,\|B(t_1)\|\, \exp\left(d_1 \int_{t_0}^{t_1} \|B\|\, dt\right) e^{\text{Re}\,(\lambda_k)t_1}\, dt_1$$
$$\leq kc_1 \int_{t_0}^t e^{\text{Re}\,(\lambda_k)(t-t_1)} e^{\text{Re}\,(\lambda_k)t_1} \|B(t_1)\| \exp\left[d_1 \int_{t_0}^{t_1} \|B(t_2)\|\, dt_2\right] dt_1$$
$$\leq \frac{kc_1}{d_1} e^{\text{Re}\,(\lambda_k)t} \exp\left[d_1 \int_{t_0}^t \|B(t_1)\|\, dt_1\right]$$

Now consider the second integral in (12). It is bounded by

$$(14) \quad c_1(n - k) \int_t^\infty e^{\mathrm{Re}\,(\lambda_n)\,(t-t_1)} \|B(t_1)\| \exp\left[\mathrm{Re}\,(\lambda_k)t_1 + d_1 \int_{t_0}^{t_1} \|B\|\,dt_2\right] dt_1$$

To obtain an upper bound, we use the following lemma:

Lemma 2. *Let a, b, and $\phi(t)$ be positive quantities satisfying the condition*

$$(15) \qquad\qquad \left|\frac{\phi'(t)}{\phi(t)}\right| \leq b < a$$

Then

$$(16) \qquad\qquad \int_t^\infty e^{-at}\phi(t)\,dt \leq \frac{e^{-at}\phi(t)}{a - b}$$

Proof

$$(17) \qquad ae^{-at}\phi(t) - e^{-at}\phi'(t) = ae^{-at}\phi(t)\,(1 - \phi'/a\phi)$$
$$\geq a\left(1 - \frac{b}{a}\right)e^{-at}$$

Thus,

$$(18) \quad e^{-at}\phi(t) = \int_t^\infty [ae^{-at}\phi(t) - e^{-at}\phi'(t)]\,dt \geq (a - b)\int_t^\infty e^{-at}\phi(t)\,dt$$

To put (14) into a form suitable for application of the lemma, we integrate by parts, obtaining

$$(19) \quad \frac{c_1(n - k)}{d_1}\,e^{\mathrm{Re}\,(\lambda_n)t} \int_t^\infty [\mathrm{Re}\,(\lambda_n - \lambda_k)]e^{-\mathrm{Re}\,(\lambda_n-\lambda_k)t_1}$$
$$\exp\left[d_1 \int_{t_0}^{t_1} \|B(t)\|\,dt\right] dt_1$$
$$+ \frac{c_1(n - k)}{d_1}\,e^{\mathrm{Re}\,(\lambda_n)t} \left[\exp\left(-\,\mathrm{Re}\,(\lambda_n - \lambda_k)t_1 + d_1 \int_{t_0}^{t_1} \|B(t)\|\,dt\right)\right]_t^\infty$$

In applying Lemma 2, we set $a = \mathrm{Re}\,(\lambda_n - \lambda_k) > 0$, and

$$\frac{\phi'(t)}{\phi(t)} = d_1\|B(t)\|$$

Since $\|B\| \to 0$ as $t \to \infty$, for t sufficiently large, and $t \geq t_0$, we will have $d_1 \sup_{t \geq t_0} \|B\| \leq \tfrac{1}{2}a = b$. Hence, using the lemma, we see that the integral in (19) is less than

$$(20) \qquad\qquad 2\,\frac{e^{-\mathrm{Re}\,(\lambda_n-\lambda_k)t}}{\mathrm{Re}\,(\lambda_n - \lambda_k)} \exp\left[d_1 \int_{t_0}^t \|B(t_1)\|\,dt_1\right]$$

Collecting terms, we obtain as a bound for the second integral in (12), the expression

$$(21) \qquad \frac{3c_1(n-k)}{d_1} \exp\left[\mathrm{Re}\,(\lambda_k)t + d_1 \int_{t_0}^{t} \|B\|\,dt \right]$$

Combining the bounds for the first and second integrals, we see that, if we choose $c_1 = 2$ and d_1 large enough so that

$$(22) \qquad 1 + \frac{2n}{d_1} + \frac{6(n-k)}{d_1} \le 2$$

then (11) will be satisfied. Once d_1 has been chosen, t_0 is chosen to satisfy $d_1 \sup\limits_{t \ge t_0} \|B\| \le b$. Thus the induction holds.

To show the convergence of z_n, we consider, as customary, the series $\sum\limits_{n=0}^{\infty} (z_{n+1} - z_n)$. There is no difficulty in showing that

$$(23) \qquad \|z_{n+1} - z_n\| \le (c_1 \sup_{t \ge t_0} \|B\|)^{n+1} e^{(\mathrm{Re}(\lambda_k) + \epsilon)t}$$

with $c_1 = c_1(\epsilon)$. Thus, if t_0 is sufficiently large, the series converges.

Let $z^{(1)}, z^{(2)}, \ldots, z^{(n)}$ be the solutions corresponding to the different characteristic roots $\lambda_1, \lambda_2, \ldots, \lambda_n$. If two roots λ_k and λ_{k+1} have the same real parts, we leave Y_1 and Y_2 unchanged and merely use the appropriate y_k.

Let us show that these n solutions are linearly independent. If these solutions were dependent, there would exist a relationship of the form

$$(24) \quad 0 = \sum_{k=1}^{n} \mu_k z^{(k)} = \sum_{k=1}^{n} \mu_k y^{(k)} + \int_{t_0}^{t} \left[\sum_{k=1}^{n} \mu_k Y_{1k}(t-t_1)B(t_1)z_k \right] dt_1$$
$$- \int_{t}^{\infty} \left[\sum_{k=1}^{n} \mu_k Y_{2k}(t-t_1)B(t_1)z_k \right] dt_1$$

Replacing Y_{1k} by $Y - Y_{2k}$, this becomes

$$(25) \qquad 0 = \sum_{k} \mu_k y^{(k)} - \int_{t_0}^{\infty} \left[\sum_{k=1}^{n} \mu_k Y_{2k}(t-t_1)B(t_1)z_k \right] dt_1$$

Now the matrix Y_{2k}, with $k = 1, 2, \ldots, n$, contains no term in $e^{\lambda_1 t}$; or if there are several λ_i, with $i = 1, 2, \ldots, l$, for which

$$\mathrm{Re}\,(\lambda_i) \le \mathrm{Re}\,(\lambda_1)$$

it contains no term with Re $(\lambda) \leq$ Re (λ_1). Therefore, since the λ_i are distinct, the terms involving exponentials with Re $(\lambda) \leq$ Re (λ_1) must vanish identically. Hence $\mu_1 = \mu_2 = \cdots = \mu_l = 0$. This condition eliminates Y_{2l}, and we are then led to the same conclusion concerning the terms involving λ_{l+1}, and so on. Thus all the $\mu_k = 0$, and we have linear independence.

We come now to the proof of the left-hand inequality of (3). Let Z be the matrix whose columns are the $z^{(k)}$. Z is nonsingular and satisfies the differential equation of (1), and in consequence its inverse $W = Z^{-1}$ satisfies the adjoint matrix equation

$$(26) \qquad \frac{dW}{dt} = -W(A + B)$$

Consider the corresponding vector equation for a row of W

$$(27) \qquad \frac{dw}{dt} = -w(A + B)$$

The same method used for the original equation shows that, for each characteristic root $-\lambda_k$ of $-A$, there is a solution $w^{(k)}$ satisfying the inequality

$$(28) \qquad \|w^{(k)}\| \leq c_2 \exp\left[- \text{Re } (\lambda_k)t + d_2 \int_{t_0}^{t} \|B\| \, dt_1 \right]$$

We can readily choose the t_0 to be the same as in the previous case.

Let V be the matrix whose rows are the $w^{(k)}$. V is nonsingular, and from the uniqueness theorem it follows that $V = DW$, where D is a constant matrix $D = (d_{ij})$. We have $VZ = DWZ = D$. Let the vector inner product of the row vector y by the column vector z be defined to be $y_1z_1 + y_2z_2 + \cdots + y_nz_n$ and be denoted by $y \cdot z$. We have

$$(29) \qquad w^{(k)} \cdot z^{(k)} = d_{kk}$$

If $d_{kk} \neq 0$, we obtain, from $|d_{kk}| \leq \|w^{(k)}\|\|z^{(k)}\|$, the inequality

$$(30) \qquad \|z^{(k)}\| \geq \left| \frac{d_{kk}}{c^2} \right| \exp\left[\text{Re } (\lambda_k)t - d_2 \int_{t_0}^{t} \|B(t_1)\| \, dt_1 \right]$$

If $d_{kk} = 0$, we proceed as follows. In place of $z^{(k)}$ we consider

$$\bar{z}^{(k)} = z^{(k)} + \sum_{l=1}^{k+1} \alpha_l z^{(l)}$$

With a change of constants, we obtain an upper bound for $\bar{z}^{(k)}$ of the type (11). The matrix $\bar{Z}$ whose columns are the $\bar{z}^{(k)}$ has the same determinant as Z and is thus nonsingular. Hence $\bar{Z} = ZE$, where E is a constant, nonsingular matrix. Similarly, in place of $w^{(k)}$, we may consider $\bar{w}^{(k)} = w^{(k)} + \sum_{l=k+1}^{n} \beta_l w^{(l)}$. Let $\bar{W}$ be the corresponding matrix, $\bar{W} = FW$. Thus $\bar{W}\bar{Z} = FWZE = FE = G$, where $G = (g_{ij})$. Consider the linear manifold M_1, composed of solutions of the adjoint equation (27) which have the property that the inner product

$$(31) \qquad w \cdot \left(z^{(k)} + \sum_{l=1}^{k-1} \alpha_l z^{(l)}\right) = 0$$

for all α_l. This is equivalent to the k independent conditions

$$(32) \qquad w \cdot z^{(1)} = w \cdot z^{(2)} = \cdots = w \cdot z^{(k)} = 0$$

Since the linear manifold M of all solutions of (27) is n-dimensional, M_1 is an $(n-k)$-dimensional manifold.

The vectors $w^{(k)} + \sum_{l=k+1}^{n} \beta_l w^{(l)}$ cannot belong to an $(n-k)$-dimensional manifold for all choices of β_l without contradicting the linear independence of the $w^{(k)}$. Consequently, for any particular k, g_{kk} cannot be zero for all choices of α_l and β_l, and therefore for each k there exists a solution of the requisite type.

This completes the proof of the first part of the theorem. It requires only minor modifications to prove the more precise result obtainable when $\int^{\infty} \|B\|\, dt < \infty$. We shall not give the details, since we shall derive a much more general result in the succeeding section.

7. Asymptotic Results. We shall now prove a more precise asymptotic result:

Theorem 8. *Let*

$$(1) \qquad \frac{dz}{dt} = (A + \phi(t) + B(t))z$$

where

(2) (a) *A is a constant matrix with simple characteristic roots λ_i*

   (b) *$\phi \to 0$ as $t \to \infty$, and $\int^{\infty} \|d\phi/dt\|\, dt < \infty$*

   (c) *$\int^{\infty} \|B(t)\|\, dt < \infty$*

   (d) *The characteristic roots $\lambda_i(t)$ of $A + \phi(t)$ either have distinct real parts or satisfy one of the following conditions:*

(3) (a) $\displaystyle \limsup_{t \to \infty} \left| \int_{t_0}^{t} \mathrm{Re}\,(\lambda_i(t) - \lambda_j(t))\,dt \right| < \infty$

   (b) $\displaystyle \lim_{t \to \infty} \int_{t_0}^{t} \mathrm{Re}\,(\lambda_i(t) - \lambda_j(t))\,dt = \infty,$

$$\int_{t_1}^{t} \mathrm{Re}\,(\lambda_i(t) - \lambda_j(t))\,dt > -c, \text{ with } t \geq t_1$$

   (c) $\displaystyle \lim_{t \to \infty} \int_{t_0}^{t} \mathrm{Re}\,(\lambda_i(t) - \lambda_j(t))\,dt = -\infty,$

$$\int_{t_1}^{t} \mathrm{Re}\,(\lambda_i(t) - \lambda_j(t))\,dt < c, \text{ with } t \geq t_1$$

Then there exist n independent solutions of (1), $x^{(k)}(t)$, *with* $1 \leq k \leq n$, *such that, as* $t \to \infty$,

(4) $$x^{(k)}(t) = \left(\exp \int_{t_0}^{t} \lambda_k(t_1)\,dt_1 \right) (c_k + o(1))$$

where c_k *is a constant, nonzero vector.*

Proof. Applying Theorem 9 of Chap. 1, we know that for $t \geq t_1$ there exists a matrix $S(t)$ such that

(5) $$S(A + \phi)S^{-1} = \Lambda$$

where Λ is a diagonal matrix with diagonal elements $\lambda_i(t)$, the characteristic roots of $A + \phi(t)$. Furthermore

(6) (a) $\displaystyle \lim_{t \to \infty} S(t) = T,\ (\det T \neq 0)$

   (b) $\displaystyle \lim_{t \to \infty} \lambda_i(t) = \mu_i$, a characteristic root of A

   (c) $\displaystyle \int^{\infty} \|dS/dt\|\,dt < \infty$

Let us perform the change of variable $y = Sz$, obtaining

(7) $$\frac{dy}{dt} = \Lambda y + \left(SBS^{-1} + \frac{dS}{dt}S^{-1} \right) y$$

Since S approaches a constant nonsingular matrix, we have

(8) $$\int^{\infty} \left\| SBS^{-1} + \frac{dS}{dt}S^{-1} \right\| dt \leq c_1 \int^{\infty} \left(\|B\| + \left\| \frac{dS}{dt} \right\| \right) dt < \infty$$

Let us introduce a new matrix $R = SBS + (dS/dt)S^{-1}$ and rewrite (7) in terms of the individual components,

(9) $$\frac{dy_i}{dt} - \lambda_i(t)y_i = \sum_{j=1}^{n} r_{ij}(t)t_j, \qquad R(t) = (r_{ij}(t))$$

For k fixed and i variable, consider the difference Re $(\lambda_i(t) - \lambda_k(t))$. Let us denote as I the set of integers i, where $0 < i \leq n$, for which this difference satisfies (3a) or (3b); the remaining set of integers where the difference satisfies (3c) we shall denote as II.

Each of the equations of (9) may be converted into an integral equation, and we do this, using two types, depending upon whether i belongs to I or II: If $i \in$ I,

$$(10) \quad y_i(t) = \delta_{ik} \exp \left(\int_a^t \lambda_k(t_1)\, dt_1 \right)$$
$$- \int_t^\infty \left[\exp \left(\int_{t_1}^t \lambda_i(s)\, ds \right) \left(\sum_{j=1}^n r_{ij}(t_1)y_j(t_1) \right) \right] dt_1$$

(δ_{ik} is the Kronecker delta symbol; δ_{ik} equals 1 if $i = k$ and equals 0 otherwise.) If $i \in$ II,

$$y_i(t) = \int_a^t \left[\exp \left(\int_{t_1}^t \lambda_i(s)\, ds \right) \left(\sum_{j=1}^n r_{ij}(t_1)y_j(t_1) \right) \right] dt_1$$

The procedure is similar to that followed in the previous proof, where we were also attempting to single out solutions with a particular growth as $t \to \infty$. To show the existence of solutions and to obtain bounds for the solutions, we use the method of successive approximations:

$$(11) \quad y_i^{(0)}(t) = \delta_{ik} \exp \left(\int_a^t \lambda_k(t_1)\, dt_1 \right), \qquad i = 1, 2, \ldots, n$$
$$y_i^{(m+1)}(t) = \delta_{ik} \exp \left[\int_a^t \lambda_k(t_1)\, dt_1 \right]$$
$$- \int_t^\infty \left[\exp \left(\int_{t_1}^t \lambda_i(s)\, ds \right) \left(\sum_{j=1}^n r_{ij}y_j^{(m)} \right) \right] dt_1, \qquad m \geq 0, (i \in \text{I})$$
$$= \int_a^t \left[\exp \left(\int_{t_1}^t \lambda_i(t_1)\, dt_1 \right) \left(\sum_{j=1}^n r_{ij}y_j^{(m)} \right) \right] dt_1, \qquad (i \in \text{II})$$

Since the proof is simple conceptually, but complicated in detail, we suggest that the reader first carry through the proof for a second-order system where the technical foliage is a minimum.

For convenience of notation, set

$$(12) \qquad \text{Re} \left(\int_a^t \lambda_i(t)\, dt \right) = H_i(t)$$
$$\Delta y_i^{(0)}(t) = |y_i^{(0)}(t)|$$
$$\Delta y_i^{(m+1)}(t) = |y_i^{(m+1)}(t) - y_i^{(m)}(t)|, \qquad m \geq 0$$

From the preceding equations we obtain, for $m \geq 1$,

$$(13) \quad \Delta y_i^{(m)}(t) \leq \int_t^\infty e^{H_i(t) - H_i(t_1)} \Big(\sum_j |r_{ij}(t_1)|\Big) \Delta y_i^{(m-1)}(t_1)\, dt_1, \qquad (i \in \mathrm{I})$$

$$\leq \int_a^t e^{H_i(t) - H_i(t_1)} \Big(\sum_j |r_{ij}(t_1)|\Big) \Delta y_i^{(m-1)}(t_1)\, dt_1, \qquad (i \in \mathrm{II})$$

and

$$(14) \qquad \Delta y_i^{(0)}(t) \leq e^{H_k(t)}$$

Let us now show by induction that a may be chosen large enough so that

$$(15) \qquad \Delta y_i^{(m)}(t) \leq 2^{-m} e^{H_k(t)}$$

The result is certainly true for $m = 0$. From (13), we obtain, using the inductive hypothesis, for $i \in \mathrm{I}$,

$$(16) \quad \Delta y_i^{(m+1)}(t) \leq 2^{-m} e^{H_k(t)} \int_t^\infty e^{H_i(t) - H_k(t) + H_k(t_1) - H_i(t_1)} \Big(\sum_j |r_{ij}(t_1)|\Big) dt_1$$

and, for $i \in \mathrm{II}$,

$$\Delta y_i^{(m+1)}(t) \leq 2^{-m} e^{H_k(t)} \int_a^t e^{H_i(t) - H_k(t) + H_k(t_1) - H_i(t_1)} \Big(\sum_j |r_{ij}(t_1)|\Big) dt_1$$

From the definition of the set I, it follows that, for $i \in \mathrm{I}$,

$$(17) \qquad \exp\Big\{ - \int_t^{t_1} \mathrm{Re}\,[\lambda_i(s) - \lambda_k(s)]\, ds \Big\} \leq c_1, \qquad t \geq t_1$$

for some fixed constant c_1. Hence for $i \in \mathrm{I}$,

$$(18) \qquad \Delta y_i^{(m+1)}(t) \leq c_1 2^{-m} e^{H_k(t)} \int_t^\infty \Big(\sum_j |r_{ij}(t_1)|\Big) dt_1$$

Similarly, since for $i \in \mathrm{II}$ we have

$$(19) \qquad \exp\Big\{ \int_{t_1}^t \mathrm{Re}\,[\lambda_i(s) - \lambda_k(s)]\, ds \Big\} \leq c_1, \qquad t \geq t_1$$

where we may use the same constant c_1 in each case, we derive

$$(20) \qquad \Delta y_i^{(m+1)}(t) \leq c_1 2^{-m} e^{H_k(t)} \int_a^t \Big(\sum_j |r_{ij}(t_1)|\Big) dt_1$$

If a is chosen so that

$$(21) \qquad \int_a^\infty \sum_{ij=1}^n |r_{ij}(t_1)|\, dt_1 \leq \tfrac{1}{2}$$

the inequality of (15) will be satisfied with m replaced by $m + 1$.

Consequently the series $\displaystyle\sum_{m=0}^{\infty} (y_i^{(m+1)} - y_i^{(m)})$ converges for each i, uniformly in any fixed t interval, and thus $y_i^{(m)} \to y_i$, where $1 \leq i \leq n$, which constitutes a solution to (10). Furthermore, for each i,

$$(22) \qquad |y_i(t)| \leq 2e^{H_k(t)}$$

Returning to (10), we see that, for $i \in \mathrm{I}$,

$$(23) \quad \left| \int_t^{\infty} \exp\left(\int_{t_1}^t \lambda_i(s)\, ds \right) \left(\sum_j r_{ij}(t_1) y_j(t_1) \right) dt_1 \right|$$
$$\leq 2c_1 e^{H_k(t)} \int_t^{\infty} \left| \sum_j r_{ij}(t_1) \right| dt_1 = o(e^{H_k(t)})$$

as $t \to \infty$. Hence for $i \in \mathrm{I}$,

$$(24) \qquad y_i(t) = (\delta_{ik} + o(1)) \exp\left(\int_a^t \lambda_k(t_1)\, dt_1 \right)$$

If $i \in \mathrm{II}$, we have

$$(25) \quad \left| \int_a^t \exp\left(\int_{t_1}^t \lambda_i(s)\, ds \right) \left(\sum_j r_{ij}(t_1) y_j(t_1) \right) dt_1 \right| \leq \left| \int_a^{t^*} \right| + \left| \int_{t^*}^t \right|$$

for $a \leq t^* \leq t$, where we shall choose a convenient t^* in a moment.
The second integral is bounded by

$$(26) \qquad 2c_1 e^{H_k(t)} \int_{t^*}^t \left(\sum_j |r_{ij}(t_1)| \right) dt_1$$

Since by hypothesis $i \in \mathrm{II}$, we have

$$(27) \qquad \int_a^t d_{ik}(s)\, ds = \int_a^t \mathrm{Re}\,[\lambda_i(s) - \lambda_k(s)]\, ds \to -\infty$$

as $t \to \infty$. Hence we may determine $t^* = t^*(t)$ such that $t^* \to \infty$ as $t \to \infty$, and

$$(28) \qquad \int_{t^*}^t d_{ik}(s)\, ds \to -\infty$$

For example, if

$$(29) \qquad F(t) = -\int_a^t d_{ik}(s)\, ds$$

choose t^* satisfying $F(t^*) = \frac{1}{2}F(t)$, $F(t_1) \leq \frac{1}{2}F(t)$, for $a \leq t_1 \leq t^*$. With this choice of t^*, the second integral is clearly $o(e^{H_k(t)})$. The first integral is bounded by

$$(30) \qquad 2c_1 e^{H_k(t)} \int_a^{t^*} \exp \left\{ \int_{t_1}^t \mathrm{Re}\,[\lambda_i(s) - \lambda_k(s)]\,ds \right\} \left(\sum_j |r_{ij}(t_1)| \right) dt$$

which again is $o(e^{H_k(t)})$.

Thus for each k we have a solution $y^{(k)}(t)$ whose components satisfy the asymptotic relations

$$(31) \qquad y_i^{(k)}(t) = (\delta_{ik} + o(1)) \exp \left(\int_a^t \lambda_k(s)\,ds \right)$$

From this it follows that the $y^{(k)}$ constitute an independent set of vectors. Since $z = S^{-1}y$ and since S tends to a constant, nonzero matrix, we have the set of vector solutions whose existence was claimed in the theorem.

8. Asymptotic Series. Frequently, far more is known about the coefficient matrix than the few properties required in the hypotheses of the preceding theorems. It is reasonable to suspect that in these cases we can learn correspondingly more about the solutions.

The impetus to our study is given by the important subclass of equations whose coefficient matrices have rational functions as elements. In this introductory discussion we assume that each of the elements in the matrix approaches a constant as $t \to \infty$. Each element then has, for large enough t, a convergent power-series expansion of the form

$$(1) \qquad a_{ij}(t) = c_0 + \frac{c_1}{t} + \cdots + \frac{c_k}{t^k} + \cdots, \qquad (c_k \equiv c_k(i,j))$$

We may then write the matrix A in the form

$$(2) \qquad A(t) = A_0 + \frac{A_1}{t} + \cdots + \frac{A_k}{t^k} + \cdots$$

where the A_k are constant matrices, for $t \geq t_0$.

Going one step further, let us consider the class of equations

$$(3) \qquad \frac{dy}{dt} = A(t)y$$

where $A(t)$ has an expansion of this above type for t sufficiently large. If (2) holds, we have

$$(4) \qquad \lim_{t \to \infty} A(t) = A_0$$
$$\lim_{t \to \infty} t(A(t) - A_0) = A_1$$
$$\cdot$$
$$\cdot$$
$$\cdot$$
$$\lim_{t \to \infty} t^{n+1} \left(A(t) - A_0 - \frac{A_1}{t} - \cdots - \frac{A_n}{t^n} \right) = A_{n+1}$$

Note that the relations of (4) in no way presuppose the convergence of (2). As a matter of fact, it is very easy to present an example of a function satisfying (4) for which the series on the right (2) is divergent for all t. Consider the scalar function

$$(5) \qquad g(t) = \int_0^\infty \frac{e^{-x}}{x + t}\, dx$$

It is easily verified that

$$(6) \qquad \lim_{t \to \infty} g(t) = 0$$
$$\lim_{t \to \infty} tg(t) = 1$$
$$\lim_{t \to \infty} t^{n+1}\left[g(t) - \frac{1}{t} + \cdots + \frac{(-1)^{n-1}(n-1)!}{t^n} \right] = (-1)^n n!$$

and that the series

$$(7) \qquad S(t) = \frac{1}{t} - \frac{1}{t^2} + \cdots + \frac{(-1)^n n!}{t^{n+1}} + \cdots$$

diverges for all values of t. Notice that the series is what one obtains formally by writing

$$(8) \qquad g(t) = \int_0^\infty \frac{e^{-x}\, dx}{t[1 + (x/t)]}$$
$$= \int_0^\infty \frac{e^{-x}}{t}\left[1 - \frac{x}{t} + \cdots + \frac{(-1)^n x^n}{t^n} + \cdots \right] dx$$

and integrating term by term—most of which is illegal.

It follows from (6) that, despite the fact that the series diverges for all t, we have, for $t \geq t_0(\epsilon)$,

$$(9) \qquad \left| g(t) - \frac{1}{t} + \cdots - (-1)^{n-1}\frac{(n-1)!}{t^n} \right| \leq \frac{n!(1 + \epsilon)}{t^{n+1}}$$

Since $n!(1 + \epsilon)/t^{n+1} \to 0$ as $t \to \infty$, we observe the startling fact that, although $S(t)$ diverges, suitable partial sums yield excellent approximations to $g(t)$ as $t \to \infty$. If, for example, $t = 10$, we obtain our best approximation by taking $n = 10$. From Stirling's formula,

$$(10) \qquad 10! \cong 10^{10}e^{-10}\sqrt{20\pi}$$

whence we see that the error term is $\cong e^{-10}\sqrt{\pi/5}$.

Series of this nature, possessing the property of furnishing good approximations to a given function if cut off at a proper stage, even if divergent as a whole, are called *asymptotic series*. Let us now give a precise definition.

Definition. *If the infinite sequence $\{a_k\}$, $k = 0, 1, 2, \ldots$, is determined as follows:*

$$\lim_{t \to \infty} f(t) = a_0$$
$$\lim_{t \to \infty} t(f(t) - a_0) = a_1$$

(11)

$$\cdot$$
$$\cdot$$
$$\cdot$$

$$\lim_{t \to \infty} t^{n+1}\left(f(t) - a_0 - \frac{a_1}{t} - \cdots - \frac{a_n}{t^n}\right) = a_{n+1}$$

then $f(t)$ is said to possess an asymptotic development as $t \to \infty$, and we write

$$(12) \qquad f(t) \sim \sum_{n=0}^{\infty} \frac{a_n}{t^n}$$

The series is not assumed convergent, and it may or may not converge.

Let us now investigate the algebra of this new correspondence we have defined between a function and a formal infinite series. We have

Theorem 9. *If*

$$(13) \qquad f(t) \sim \sum_{n=0}^{\infty} a_n t^{-n}, \qquad g(t) \sim \sum_{n=0}^{\infty} b_n t^{-n}$$

then, for any two constants c_1 and c_2, we have

$$(14) \qquad c_1 f + c_2 g \sim \sum_{n=0}^{\infty} (c_1 a_n + c_2 b_n) t^{-n}$$

Furthermore,

$$(15) \qquad fg \sim \sum_{n=0}^{\infty} c_n t^{-n}$$

where

$$(16) \qquad c_n = \sum_{k+l=n} a_k b_l$$

Moreover, if $a_0 \neq 0$,

$$(17) \qquad \frac{1}{f(t)} \sim c_0 + \frac{c_1}{t} + \cdots + \frac{c_n}{t^n} + \cdots$$

where

$$(18) \quad a_0 c_0 = 1, \ a_0 c_1 + a_1 c_0 = 0, \ \ldots, \ \sum_{k+l=n} a_k c_l = 0, \qquad n > 0$$

Finally, if

$$(19) \qquad\qquad f'(t) \sim \sum_{n=2} d_n t^{-n}$$

then $d_n = -(n-1)a_{n-1}$. *If* $a_0 = a_1 = 0$, *then*

$$(20) \qquad\qquad \int_t^\infty f(t)\, dt \sim \sum_{n=2}^\infty \frac{a_n t^{-(n-1)}}{(n-1)}$$

The proofs follow immediately from the definition, and we leave them as exercises.

We may summarize the above by noting that asymptotic series may be handled, as far as algebraic properties are concerned, like ordinary power series. The same holds for the operations of differentiation and integration, provided that the resultant functions possess asymptotic developments. Hence, if $f, f', \ldots, f^{(n)}$ all possess asymptotic developments, then $P(f, f', \ldots, f^{(n)})$, where P is any polynomial, possesses the asymptotic development that one computes formally.

To show that it is not generally true that f' has an asymptotic development if f has, consider the function

$$(21) \qquad\qquad f = \frac{1}{t} + e^{-t} \sin e^{2t}$$

According to our definition, $f \sim 1/t$. However,

$$(22) \qquad\qquad f' = -\frac{1}{t^2} - e^{-t} \sin e^{2t} + 2e^t \cos e^{2t}$$

is *not* asymptotic to $-t^{-2}$.

This example also illustrates the fact that, although a function possesses a unique asymptotic development, many functions may possess the same asymptotic development.

Let us now turn to the application of the preceding ideas to the theory of differential equations. Taking a simple equation, such as

$$(23) \qquad\qquad u'' - \left(1 + \frac{1}{t^2}\right) u = 0$$

we see, according to Theorem 8, that the solutions are linear combinations of functions asymptotic to e^t and e^{-t} as $t \to \infty$. Let us try a solution of the form

$$(24) \qquad u = e^t \left(1 + \frac{c_1}{t} + \frac{c_2}{t^2} + \cdots + \frac{c_n}{t^n} + \cdots \right)$$

Substituting into (23) and equating coefficients, we obtain

$$(25) \qquad c_1 = \tfrac{1}{2}, \qquad c_n = \frac{(n^2 - n - 1)}{2n} c_{n-1}, \qquad n \geq 2$$

Consequently the series in (24) diverges for all t. We shall show below, however, that this series we have found formally is actually an asymptotic series for a solution of (23). In view of the form of (24), it will be useful to liberalize slightly our definition of an asymptotic development and write

$$(26) \qquad f(t) \sim \phi(t) \sum_{n=0}^{\infty} a_n t^{-n}$$

if

$$(27) \qquad \left| f(t) - \phi(t) \sum_{n=0}^{m} a_n t^{-n} \right| \leq |\phi(t)| c_m t^{-(m+1)}, \qquad m = 0, 1, 2, \ldots,$$

with c_m a constant, as $t \to \infty$. If $\phi(t) \geq b > 0$ for $t \geq t_0$, we may write

$$(28) \qquad \frac{f(t)}{\phi(t)} \sim \sum_{n=0}^{\infty} a_n t^{-n}$$

Let us observe that, once we know that we have an asymptotic series for a solution of (23), it is easy to compute the coefficients $a_1, a_2, \ldots,$ $a_n, \ldots$ recursively by equating coefficients. The main problem in connection with the application of asymptotic series to differential equations is the following:

Given an infinite series of the form $\sum_{n=0}^{\infty} a_n t^{-n}$, *divergent for all t, formally satisfying the differential equation* $P(u, u', u'', \ldots, u^{(n)}) = 0$, *under what conditions is the series the asymptotic series of a solution of the differential equation?*

We shall give a partial answer to the problem for the case where P is a linear homogeneous form in $u, u', u^{(2)}, \ldots, u^{(n)}$, in which case $P = 0$ is a linear differential equation

$$(29) \qquad u^{(n)} + a_1(t)u^{(n-1)} + \cdots + a_n(t)u = 0$$

As we know, the theory of equations of this type can be made part of the theory of linear systems,

$$(30) \qquad \frac{dy}{dt} = A(t)y$$

and it is in this form that we shall treat it.

9. The Asymptotic Behavior of the Solution of $dy/dt = A(t)y$. The problem has been solved completely of determining the asymptotic behavior of the solutions of

$$(1) \qquad \frac{dy}{dt} = A(t)y$$

where the elements of $A(t)$ are rational functions of t, or more generally, possess asymptotic developments of the form

$$(2) \qquad a_{ij}(t) \sim p(t) + \sum_{n=1}^{\infty} c_n t^{-n}, \qquad p \equiv p_{ij}, c_n = c_n(i,j)$$

where $p(t)$ is a polynomial in t. However, the solution is quite complicated in detail, although not in principle, and for that reason we shall discuss only the simple case where $A(t)$ has the asymptotic expansion

$$(3) \qquad A(t) \sim A_0 + A_1 t^{-1} + \cdots + A_n t^{-n} + \cdots$$

Furthermore, we shall assume that the characteristic roots of A_0 are distinct. It is easy to give examples illustrating the complicated nature of the solutions in the case where A_0 has multiple characteristic roots. For example, as we shall see in Chap. 6, there are two solutions of

$$(4) \qquad u'' - \frac{u}{t} = 0$$

having asymptotic developments of the form

$$(5) \qquad u \sim (\exp \pm 2\sqrt{t})t^{c_0}(1 + c_1 t^{-1} + \cdots + c_n t^{-n} + \cdots)$$

Exercise

1. Determine c_0 by equating coefficients.

There are many special techniques particularly applicable to the second-order equation; hence a complete discussion of the nature of the solution is much easier than for the general nth-order equation.

Our principal result is

Theorem 10. *Consider* (1), *where*

(6) (a) $A(t) \sim A_0 + A_1 t^{-1} + \cdots + A_n t^{-n} + \cdots$
        (b) *The characteristic roots* $\lambda_1, \lambda_2, \ldots, \lambda_n$ *of* A_0 *are simple*

Then, corresponding to any particular characteristic root λ_k, there is a solution y_k of (1) possessing the asymptotic development

$$(7) \qquad y_k \sim e^{\lambda_k t} t^{\mu_k} \left(c_0 + \frac{c_1}{t} + \cdots + \frac{c_n}{t^n} + \cdots \right)$$

where c_0 is a nontrivial vector.

Proof. We require first the following

Lemma 3. *If each coefficient of the algebraic equation*

$$(8) \qquad f(z) = z^n + a_1(t)z^{n-1} + \cdots + a_n(t) = 0$$

possesses an asymptotic expansion

$$(9) \qquad a_i(t) \sim \sum_{k=0}^{\infty} c_k^{(i)} t^{-k}$$

and if the equation

$$(10) \qquad g(z) = z^n + c_0^{(1)} z^{n-1} + \cdots + c_0^{(n)} = 0$$

possesses simple characteristic roots $r_1, r_2, \ldots, r_n$, then each root of (8) has the asymptotic expansion

$$(11) \qquad r_i(t) \sim r_i + \sum_{k=1}^{\infty} b_k^{(i)} t^{-k}$$

Proof of Lemma. It is convenient, as before, to carry through the proof using complex-variable methods. Since (10) has simple roots, the same will be true of (8) for t large. For $t \geq t_0$ then, we may draw small circles C_i around each root r_i in the complex plane, which are nonintersecting. Applying Cauchy's residue theorem, it follows that

$$(12) \quad r_i(t) = \frac{1}{2\pi i} \int_{C_i} \frac{z f'(z)}{f(z)} \, dz, \qquad i = 1, 2, \ldots, n, \; (f(z) \equiv f(z,t))$$

For each z on C_i we have, as $t \to \infty$,

$$(13) \qquad \frac{z f'(z)}{f(z)} = \frac{z g'(z)}{g(z)} + \frac{e_1(z)}{t} + \cdots + \frac{e_n(z)}{t^n} + O(t^{-(n+1)})$$

Integrating term by term, and noting that the $O(t^{-(n+1)})$ estimate holds uniformly in z, on each C_i, we obtain the required asymptotic expansion for $r_i(t)$.

Let us now turn to the proof of our theorem. The system has the form

$$(14) \qquad \frac{dz}{dt} = \left(A_0 + \frac{A_1}{t} + A_2(t) \right) z$$

where $\|A_2(t)\| \leq c_1 t^{-2}$ as $t \to \infty$. Since the characteristic roots of A_0 are distinct and $\int^{\infty} \|A_2(t)\| \, dt < \infty$, our system satisfies the conditions of Theorem 8. We know then that there exist n solutions $z_1, z_2, \ldots, z_n$ with the asymptotic forms

$$(15) \qquad z_k \sim e^{\lambda_k t} t^{\mu_k} [c_k + o(1)]$$

where c_k is a constant, nonzero vector. By means of a change of variable, we may always assume that the λ_k are distinct from zero. This is not essential, but it simplifies some of the details.

The proof is inductive, starting from the known result of (15), using the same integral equations as before. Assuming that we have shown that

$$(16) \qquad z_k = e^{\lambda_k t} t^{\mu_k} [c_k + c_k^{(1)} t^{-1} + \cdots + c_k^{(n)} t^{-n} + o(t^{-n})]$$

for $n = 0, 1, \ldots, m$ and for each k, the integral equation is used to show that the same expressions are valid for $m + 1$.

This exercise in repeated integration by parts we leave to the reader. We suggest that, before the general proof is attempted, the exercises below be worked.

Exercises

2. Derive the asymptotic expansions of the solution of

$$(17) \qquad u'' \pm (1 + g(t))u = 0$$

where

$$(18) \qquad g(t) \sim \frac{g_1}{t} + \frac{g_2}{t^2} + \cdots$$

3. Use the asymptotic expansions derived above to find asymptotic expansions for the zeros of the solutions of $u'' + (1 + g(t))u = 0$.

Miscellaneous Exercises

1. All solutions of $d^2z/dt^2 = (A + B(t) + C(t))z$ are bounded, provided that

(a) A is a constant, negative definite matrix
(b) $B(t)$ is symmetric
(c) $(1 + c_1) \left| \sum_{i,j=1}^{n} b_{ij}(t)x_i x_j \right| \leq \left| \sum_{i,j=1}^{n} a_{ij} x_i x_j \right|$ for $t \geq t_0$ and some $c_1 > 0$
(d) $\int^{\infty} \|dB/dt\| \, dt < \infty, \quad \int^{\infty} \|C(t)\| \, dt < \infty$

2. If $dy/dt = A(t)y$, $y(0) = y_0$, then $\|y\| \leq \|y_0\| \exp \left[\int_0^t \|A(t_1)\| \, dt_1. \right]$

3. If $\int^{\infty} \|A(t)\| \, dt < \infty$, $\lim\limits_{t \to \infty} y$ exists. (Trjitzinsky.)

4. If $\sum\limits_{i,j=1}^{n} \int_{0}^{\infty} |a_{ij} + a_{ji}| \, dt < \infty$, all solutions of $dy/dt = A(t)y$ are bounded as $t \to \infty$.

5. There exists an orthogonal matrix $B(t)$ such that if $y = B(t)z$, the equation $dy/dt = A(t)y$ is transformed into $dz/dt = A^*(t)z$, where $A^*(t)$ is semidiagonal. (Diliberto.)

6. There exists a bounded nonsingular matrix $B(t)$, such that $A^*(t)$ is diagonal. (Diliberto.)

BIBLIOGRAPHY

Section 2

Bellman, R., *The stability of solutions of linear differential equations*, Duke Math. J., vol. 10 (1943), pp. 643–647.

Caligo, O., *Un criterio sufficiente di stabilita* . . . , pp. 177–185, Congresso Un. Mat. Ital. Bologna, 1940.

Cesari, L., *Sulla stabilita delle soluzione delle equazioni differenziali lineari*, Annali R. Scuola Norm. Sup. Pisa, ser. 2, vol. 8 (1939), pp. 131–148.

Dini, U., *Studi sulla equazioni differenziali lineari*, Annali di Mat., ser 3. vol. 3 (1900), pp. 125–183.

Hukuwara, M., *Sur les points singuliers des équations différentielles linéaires*, J. Fac. Sci. Hokkaido Imp. Univ., ser. 1, vol. 2 (1934), pp. 13–88.

Levinson, N., *The asymptotic behavior of a system of linear differential equations*, Amer. J. Math., vol. 68 (1946), pp. 1–6.

Späth, H., *Ueber das asymptotische Verhalten des Lösungen nichthomogener linearer Differentialgleichungen*, Math. Zeit., vol. 30 (1929), pp. 487–513.

Weyl, H., *Comment on the preceding paper*, Amer. J. Math., vol. 68 (1946), pp. 7–12.

Concerning the fundamental lemma, see also:

Gronwall, T. H., *Note on the derivatives with respect to a parameter of the solutions of a system of differential equations*, Ann. of Math., vol. 20 (1918), pp. 292–296.

Guiliano, L., *Generalazzione di un lemma di Gronwall* . . . , Rend. Accad. Lincei (1946), pp. 1264–1271.

Section 3. See the above reference to L. Cesari.

Section 5. For Theorem 7 see O. Perron, Math. Ann., vol. 143 (1913), pp. 25–50. For the proof of Theorem 8, see the above references to R. Bellman, and O. Caligo, and also A. Wintner, Amer. J. Math., vol. 68 (1946), pp. 185–213.

Section 6

Bellman, R., *The boundedness of solutions of linear differential equations*, Duke Math. J., vol. 14 (1947), pp. 83–97.

For the result on asymptotic behavior of the solutions see

Dunkel, O., *Regular singular points of a system of homogeneous linear differential equations of the first order*, Proc. Amer. Acad. Arts Sci., vol. 38 (1912–1913), pp. 341–370.

Section 7

Levinson, N., *The asymptotic nature of the solutions of linear systems of differential equations*, Duke Math. J., vol. 15 (1948), pp. 111–126.

Section 8.   A classical paper on the application of asymptotic series to the theory of differential equations is that of E. Borel, *Mémoire sur les séries divergentes*, Ann. École Norm., vol. 16(1899), pp. 9–136.   See also his book, *Leçons sur les séries divergentes*, Gauthier-Villars & Cie, Paris, 1901.

Section 9.   Using a different method, this result was derived by M. Hukuwara in the paper cited above.   For a survey of results up to 1938, see:

Trjitzinsky, W. J., *Singular point problems in the theory of linear differential equations*, Bull. Amer. Math. Soc., vol. 44 (1938), pp. 209–233.

CHAPTER 3

THE EXISTENCE AND UNIQUENESS OF SOLUTIONS OF NONLINEAR SYSTEMS

1. Introduction. In previous chapters we have discussed the properties of linear systems of the form $dz/dt = Az$. We now turn to a preliminary discussion of nonlinear systems of the form

$$(1) \qquad \frac{dz_i}{dt} = f_i(z_1, z_2, \ldots, z_n, t)$$
$$z_i(0) = c_i, \qquad i = 1, 2, \ldots, n$$

Introducing the new dependent variable $z_{n+1} = t$, we may write (1) in the form

$$(2) \qquad \frac{dz_i}{dt} = f_i(z_1, z_2, \ldots, z_n, z_{n+1}), \qquad i = 1, 2, \ldots, n$$
$$\frac{dz_{n+1}}{dt} = 1$$
$$z_i(0) = c_i, \qquad i = 1, 2, \ldots, n$$
$$z_{n+1}(0) = 0$$

where the right-hand side is now free of any explicit dependence upon t. This in turn may be written in the simpler form

$$(3) \qquad \frac{dz}{dt} = f(z), \qquad z(0) = c$$

where $f(z)$ denotes the vector whose ith component is $f_i(z_1, z_2, \ldots, z_{n+1})$.

We wish to derive some simple conditions which ensure that (3) has a unique solution. More important, however, is the opportunity to display two fundamental methods, one of extreme theoretical importance, the other of extreme practical importance in connection with the numerical solution of differential equations, ordinary and partial. The first is the method of successive approximations, which we encountered previously, and the second is the method of finite differences. This last consists in replacing (3) by the difference equation

$$(4) \qquad z(t + h) - z(t) = hf(z), \qquad z(0) = c$$

where t takes only the values 0, h, $2h$, and so on.

A third powerful method for providing existence theorems, the Birkhoff-Kellogg technique of fixed points in function space, will not be discussed in this volume because of its dependence upon advanced concepts.

Since we are not primarily interested in questions of existence and uniqueness in the small, we shall content ourselves with stating and proving only the basic results.

2. Method of Successive Approximations. The natural extension of the method we employed for linear systems is to define inductively the sequence $\{z_n\}$ as follows:

$$(1) \qquad z_0 = c$$
$$\frac{dz_{n+1}}{dt} = f(z_n), \qquad z_{n+1}(0) = c, \qquad n = 0, 1, \ldots$$

This is equivalent to the integral definition

$$(2) \qquad z_0 = c$$
$$z_{n+1} = c + \int_0^t f(z_n) \, dt_1$$

Let us assume that $f(z)$ is a continuous function of z in some neighborhood of c, say the region R defined by $\|z - c\| \leq c_1$. The definition given in (2) is inductive and hence gives rise to the question of the actual existence of the z_n for $n \geq 2$, since $f(z_n)$ may fail to be defined for some n. Let us show that by restricting t to a suitable interval we can ensure the existence of each z_n. From (2) we obtain

$$(3) \qquad \|z_{n+1} - c\| \leq \int_0^t \|f(z_n)\| \, dt_1 \leq c_2 t$$

where we let $c_2 = \max \|f(z)\|$ for z in R. Hence if $c_2 t \leq c_1$, z_{n+1} will also be in R. Henceforth we restrict t to lie in the interval $0 \leq t < c_1/c_2$. Note carefully the strict inequality, which implies that z_n is always inside R.

We must now consider the question of convergence of the sequence $\{z_n\}$. As before, this is equivalent to the convergence of the series $\sum_{n=0}^{\infty} (z_{n+1} - z_n)$. In place of this series, we consider the majorant $\Sigma \|z_{n+1} - z_n\|$. From (2) we obtain for $n \geq 1$, the inequality

$$(4) \qquad \|z_{n+1} - z_n\| \leq \int_0^t \|f(z_n) - f(z_{n-1})\| \, dt$$

To continue the proof along the lines of Sec. 3 of Chap. 1, we require some relation between $\|f(z_n) - f(z_{n-1})\|$ and $\|z_n - z_{n-1}\|$. Let us assume that for any two vectors x and y in R we have the relation

(5) $$\|f(x) - f(y)\| \le c_3\|x - y\|$$

where c_3 is a constant depending only upon the region R and not upon the vectors x and y. A condition of this type is called a *Lipschitz condition*. It automatically implies continuity.

Returning to (4), and using (5), there results

(6) $$\|z_{n+1} - z_n\| \le c_3 \int_0^t \|z_n - z_{n-1}\| \, dt_1, \qquad n \ge 1$$

Since $\|z_1 - z_0\| \le \int_0^t \|f(z_0)\| \, dt_1 = \|f(c)\|t = c_4 t$, iteration of (6) yields

(7) $$\|z_{n+1} - z_n\| \le \frac{c_4 c_3^n t^{n+1}}{(n+1)!}$$

whence $\Sigma \|z_{n+1} - z_n\|$ converges uniformly for $0 \le t \le t_0 < c_1/c_2$. It follows then that z_n converges uniformly to a function $z(t)$ which satisfies the integral equation

(8) $$z = c + \int_0^t f(z) \, dt$$

and thus the differential equation.

Let us, before turning to the treatment of the uniqueness problem, mention a simple condition on $f(z)$ which will yield the Lipschitz condition. The mean-value theorem shows that, if $f(z)$ has uniformly bounded partial derivatives with respect to the z_i in R, it will satisfy a Lipschitz condition in R.

3. Uniqueness. We now wish to show that the solution found by the method of successive approximations is, under the assumptions we have made, the only solution of

(1) $$\frac{dz}{dt} = f(z), \qquad z(0) = c$$

in the interval $0 \le t \le t_0 < c_1/c_2$. Assume that there exists another solution y. Since y is continuous and in R at time $t = 0$, it is in R for $0 \le t \le t_1$, where t_1 is a positive quantity. Let $t_2 = \min [t_0, t_1]$. For $0 \le t \le t_2$ we have, combining

(2) $$y = c + \int_0^t f(y) \, dt_1$$

and (2) of the previous section, the inequality

(3) $$\|z_{n+1} - y\| \le \int_0^t \|f(z_n) - f(y)\| \, dt_1$$

and thence

$$(4) \qquad \|z_{n+1} - y\| \leq c_3 \int_0^t \|z_n - y\| \, dt_1$$

Using the fact that $\|z_0 - y\| \leq \int_0^t \|f(y)\| \, dt_1 \leq c_2 t$, we obtain by iteration

$$(5) \qquad \|z_{n+1} - y\| \leq \frac{c_2 c_3^n t^{n+1}}{(n+1)!}$$

Letting $n \to \infty$, we see that $\|z - y\| \leq 0$, which means that $z \equiv y$ in $[0, t_2]$. If $t_2 = t_0$, our proof is finished. If not, we can begin at $t = t_2$ and obtain a larger interval within which $y \equiv z$. If, however, we continue in this direct fashion, we have no guarantee that we can ever fill up the entire interval $[0, t_0]$. Hence we proceed as follows: We know that we can find a nonzero interval $[0, \tau]$ within which $z \equiv y$. Since y and z are continuous, this interval must be closed. Let $[0, \tau]$ be the largest such interval. If $\tau < t_0$, we may employ the method above to increase the interval. Hence $\tau = t_0$.

We have now completed the proof of the following result:

Theorem 1.   *If, for any two vectors x and y in the region R defined by $\|z - c\| \leq c_1$, we have*

$$(6) \qquad \|f(x) - f(y)\| \leq c_3 \|x - y\|$$

where c_3 is a constant depending only upon R, there exists a unique solution to

$$(7) \qquad \frac{dz}{dt} = f(z), \qquad z(0) = c$$

for $0 \leq t < c_1/c_2$, where $c_2 = \max_R \|f(z)\|$.

Exercises

1. Consider the sequence defined by

$$(8) \qquad \begin{aligned} z_0 &= w(t) \\ z_{n+1} &= c + \int_0^t f(z_n) \, dt_1, \qquad n = 0, 1, \ldots \end{aligned}$$

where $w(0) = c$. Does this sequence converge to the solution of the differential equation, *under the above conditions on $f(z)$* and under suitable restrictions on $w(t)$?

2. Do there exist functions $f(z)$, apart from linear functions, which satisfy Lipschitz conditions for all real z?

3. Is it necessary for the existence and uniqueness of the solution that $f(z)$ be continuous? Consider, for example, the scalar equation

(9)   (a) $du/dt = f(u)$, $u(0) = \frac{1}{4}$, where
      (b) $f(u) = 0$, $-\infty < u < \frac{1}{2}$, $= 1$, $u \geq \frac{1}{2}$

What generalization of Theorem 1 is valid?

---

4. An Example Illustrating Lack of Uniqueness. We have seen that the Lipschitz condition yields the existence and uniqueness of the solution of

$$(1) \qquad \frac{dy}{dt} = f(y), \qquad y(0) = c$$

Let us now assume only that $f(y)$ is continuous. As we shall see below, this condition is sufficient to guarantee the existence of at least one solution of (1). However, we cannot establish uniqueness on this hypothesis, since it is not true in general.

Let us consider a simple example. The scalar equation

$$(2) \qquad \frac{du}{dt} = \sqrt{u}, \qquad u(0) = 0$$

has two solutions

$$(3) \qquad u = 0$$
$$u = \frac{t^2}{4}$$

for $t \geq 0$. Of course, $\sqrt{u}$ does not satisfy the Lipschitz condition in the vicinity of $u = 0$.

Exercises

1. Show that the equation $du/dt = u^a$, $u = 0$ at $t = 0$, possesses two solutions for $0 < a < 1$, but not for $a = 0$ or 1.

2. Consider the equation $du/dt = u(\log u)^a$, $u = 0$ for $t = 0$. For what values of a is there a unique solution?

3. Consider $du/dt = f(u)$, where $u = 0$ at $t = 0$, $\int_0 du/|f(u)| = \infty$. Is the solution unique in this case?

---

5. Method of Finite Differences. Let us replace the differential equation we have been treating by the difference equation

$$(1) \qquad \frac{y(t + h) - y(t)}{h} = f(y(t)), \qquad t = 0, h, 2h, \ldots$$

The solution of (1) may be continued up to $t = nh$, provided that $y(kh)$ lies in R for $k = 0, 1, \ldots, n - 1$. As above, this is true for $nh < c_1/c_2$, where c_1 and c_2 have their previous meaning.

The geometric significance of this approximation may be illustrated very simply for the case of scalar equations. Suppose that u is a solution of $du/dt = f(u)$, $u(0) = c_1$, and we wish to find u at the point t_1. Let us set $t_1 = 3h$ and subdivide the interval $[0,t_1]$ into three equal parts of length h, as in Fig. 1. Assuming that the solution curve is a straight line over the interval $[0,h]$, we find that its equation is

$$(2) \qquad u = c + tf(0), \qquad 0 \le t \le h$$

since its slope is determined by means of the differential equation. At $t = h$, $u = c + hf(c)$. From P to Q we again assume that the curve is a straight line, determining the new slope by means of the differential equation. The equation of PQ is then

$$(3) \qquad u = u(h) + f(u(h))(t - h), \qquad h \le t \le 2h$$

where $u(h) = c + hf(c)$. Continuing in this fashion, we determine the line QR and thus $u(3h)$. If we wish a better approximation, we repeat the procedure, using six intervals instead of three, and then twelve intervals, and so on.

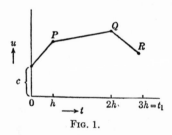

FIG. 1.

We shall show that, under the sole assumption that $f(u)$ is continuous, it is possible to obtain a sequence of solutions of the difference equation for different values of h, which approach a solution of the differential equation. This proof will be only an existence proof, nonconstructive in the sense that we cannot in this way with certainty actually compute solutions. If we wish to obtain numerical solutions, we must impose our former condition, the Lipschitz condition. We shall leave this last as an exercise and consider in detail only the case where we assume continuity alone.

For simplicity, let us set $h_1 = h$, $h_{n+1} = h_n/2 = h/2^n$. For each h_k we have another difference equation (a vector-matrix equation)

$$(4) \qquad \frac{y(t + h_k) - y(t)}{h_k} = f(y(t)), \qquad t = 0, h_k, 2h_k, \ldots$$

For each h_k, we may compute, using the difference equation, the set of values $y(0), y(h_k), \ldots, y(nh_k)$, where n is the largest integer satisfying $nh_k < c_1/c_2$. Clearly, we shall have a different n for each k. Let us now construct the function which assumes the values $y(0)$, $y(h_k)$, . . . at $t = 0, h_k, . . .$, respectively, and is linear in between. This function is shown schematically, in Fig. 2.

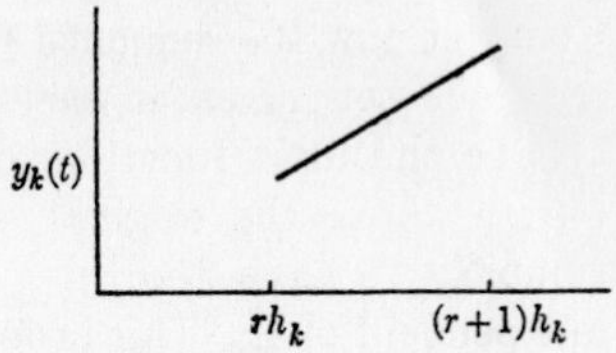

We shall demonstrate that the sequence $\{y_k(t)\}$ satisfies a uniform Lipschitz condition,

$y_k(t)$

$rh_k \qquad (r+1)h_k$

Fig 2.

$$(5) \quad |y_k(t) - y_k(s)| \leq c_2|t - s|$$

$$0 \leq t, s < \frac{c_1}{c_2}$$

where $c_2 = \max_{y \in R} \|f(y)\|$ is clearly independent of k.

This we see as follows: If s and t are within one interval, then

$$(6) \quad y_k(t) - y_k(s) = (t - s)f(y_k(rh_k))$$

since within each interval $[rh_k, (r + 1)h_k]$, we have

$$y_k(t) = y_k(rh_k) + (t - rh_k)f(y_k(rh_k))$$

If s and t are in adjacent intervals, $s < rh_k < t$, we write

$$(7) \quad y_k(t) - y_k(s) = y_k(t) - y_k(rh_k) + y_k(rh_k) - y_k(s)$$

and obtain

$$(8) \quad \begin{aligned} \|y_k(t) - y_k(s)\| &\leq \|y_k(t) - y_k(rh_k)\| + \|y_k(rh_k) - y_k(s)\| \\ &\leq c_2[(t - rh_k) + (rh_k - s)] = c_2(t - s) \end{aligned}$$

Quite generally then, if s and t are any two points in $[0,c_1]$, by writing

$$(9) \quad t - s = (t - rh_k) + (rh_k - (r - 1)h_k) + \cdots + (rh_k - s)$$

we derive the general inequality

$$(10) \quad \|y_k(t) - y_k(s)\| \leq c_2|t - s|$$

From this inequality it follows that, whenever $|t - s| \leq \delta$ in $[0,c_1]$, we have $\|y_k(t) - y_k(s)\| \leq \epsilon$, for all k, provided only that $\delta = \delta(\epsilon)$, independent of k. A sequence $\{y_k(t)\}$ satisfying this condition is said to be *equicontinuous*.

We now wish to establish a general result:

Lemma (Arzela Selection Theorem). *Let $\{y_n(t)\}$ be an infinite sequence of uniformly bounded, equicontinuous functions in a bounded interval $[a,b]$. Then there exists a subsequence which converges uniformly in $[a,b]$.*

Proof. Let $t_1, t_2, . . .$ be the rational points in $[a,b]$, enumerated in some order. The sequence $\{y_n(t_1)\}$ is uniformly bounded and hence possesses a convergent subsequence $\{y_{n1}(t_1)\}$. Let us now consider the sequence $\{y_{n1}(t_2)\}$. In turn, as a uniformly bounded sequence, it possesses a convergent subsequence $\{y_{n2}(t_2)\}$. Continuing in this fashion, we obtain a sequence $\{y_{nk}(t)\}$ which converges for $t = t_1, t_2, \ldots, t_k$. Consider now the sequence $\{y_{kk}(t)\}$. By virtue of our construction, this sequence converges at each of the points $t_1, t_2, \ldots, t_n, \ldots$. Let $y(t)$ be the limit function, defined as yet only for $t = t_1, t_2, \ldots, t_n,$ $\ldots$. Since the original sequence $\{y_n(t)\}$ was taken to be equicontinuous, we have $\|y_{kk}(t_i) - y_{kk}(t_j)\| \leq \epsilon$ for $|t_i - t_j| \leq \delta = \delta(\epsilon)$, δ being independent of k. This property, then, is preserved by the limit function $y(t)$, that is, $\|y(t_i) - y(t_j)\| \leq \epsilon$ when $|t_i - t_j| \leq \delta$, which means that $y(t)$ is continuous over the set $\{t_i\}$. We now define $y(t)$ for all t in $[a,b]$ by means of the relation $y(t) = \lim y(t_i)$ as $t_i \to t$ through a sequence of rational values.

It remains to show that $y_{kk}(t) \to y(t)$ uniformly for t in $[a,b]$. Divide $[a,b]$ into N equal parts, where N will be specified in a moment, and let the end points be $a = s_0, a_1, \ldots, s_N = b$. If $s_r \leq t \leq s_{r+1}$, we have

$$(11) \quad y_{kk}(t) - y(t) = (y_{kk}(t) - y_{kk}(s_r)) + (y_{kk}(s_r) - y(s_r)) \\ + (y(s_r) - y(t))$$

Choose N so that $\|y_{kk}(t) - y_{kk}(s_r)\| \leq \epsilon$ whenever $s_r \leq t \leq s_{r+1}$, for all k. This may be done by virtue of the equicontinuity. This also implies $\|y(s_r) - y(t)\| \leq \epsilon$.

Then choose $n \geq n_0$, depending upon N so that $\|y_{kk}(s_l) - y(s_l)\| \leq \epsilon$, for $l = 0, 1, 2, \ldots, N$. It follows that for $n \geq n_0$, depending only upon ϵ, we have $\|y_{kk}(t) - y(t)\| \leq 3\epsilon$, which demonstrates the uniform convergence.

Exercise

1. Show by means of counterexamples that the above theorem is not valid if we omit any one of the three conditions "equicontinuous," "bounded," "finite interval."

Having established this result, we apply it to the sequence $\{y_n(t)\}$ obtained from the difference equations. We must now show that the limit function satisfies the differential equation. This may be shown directly, but it is easier, as usual, to show that y satisfies the integral equation. For any l and k we have

$$(12) \qquad y_k[(1 + l)h_k] = y_k(lh_k) + h_k f[y_k(lh_k)]$$

Summing over $l = 0, 1, 2, \ldots, L$, we have

$$(13) \qquad y_k[(L + 1)h_k] = c + \sum_{l=0}^{L} h_k f[y_k(lh_k)]$$

since $y_k(0) = c$ for each k. Let L be chosen so that $Lh_k \to t$ as $k \to \infty$ and $h \to 0$; for example, $L = [2_k t/h]$. Then $y_k[(L+1)h_k] \to y(t)$. Since the right-hand sum in (13) looks like the approximating sum to a Riemann integral, we should expect this expression to approach $\int_0 f(y) \, dt$ as $k \to \infty$. We have

$$(14) \qquad \sum_{l=0}^{L} h_k f[y_k(lh_k)] = \sum_{l=0}^{L} h_k f[y(lh_k)] + \sum_{l=0}^{L} h_k \{f[y_k(lh_k)] - f[y(lh_k)]\}$$

The first sum is a Riemann sum and approaches $\int_0^t f(y) \, dt$. Since the sequence $\{y_k(t)\}$ is uniformly convergent and since $f(y)$ is continuous, we have $\|f[y_k(lh_k)] - f[y(lh_k)]\| \leq \epsilon$ for all l and for $k \geq k_0$. Hence

$$(15) \qquad \left\| \sum_{l=0}^{L} h_k \{f[y_k(lh_k)] - f[y(lh_k)]\} \right\| \leq \epsilon h_k L \leq \epsilon t$$

This completes the proof.

Exercises

2. Prove directly that $y(t)$ is differentiable and satisfies the differential equation $y' = f(y)$.

3. Under the hypothesis of Theorem 1, show that z is a continuous function of the initial vector c in some region about c.

4. Let $z(t,c)$ be the solution of $dz/dt = f(z)$, $z(0) = c$, where f satisfies the above conditions. Show that for s and $t \geq 0$ and sufficiently small, we have $z(s + t, c) = z(s, z(t,c))$.

5. Consider the scalar equation $du/dt = au + u^n$ (the Bernoulli equation). It may be solved in elementary terms by letting $v = 1/u^{n-1}$. Show directly that the above functional equation is satisfied.

6. In the fixed interval $[0,a]$, find an expression for the difference between the solutions of the two vector systems

$$y' = f(y), \qquad y(0) = c_1$$
$$z' = g(z), \qquad g(0) = c_2$$

7. (Generalization of Newton's method.) Compare the rapidity of convergence of the successive approximations obtained from

$$(16) \qquad y'_{n+1} = f(y_n), \qquad y_{n+1}(0) = c$$

with those obtained from

$$(17) \qquad y'_{n+1} = f(y_n) + (y_{n+1} - y_n)f'(y_n), \qquad y_{n+1}(0) = c$$

8. Consider the system $dy_i/dt = f_i(y_1, y_2, \ldots, y_n, t)$, $i = 1, 2,$ $\ldots, n$, where $|f_i(y,t) - f_i(z,t)| \leq \sum_{k=1}^{n} l_k |y_k - z_k|$ for all y_k and z_k. Let $(y_1, y_2, \ldots, y_n)$ and $(z_1, z_2, \ldots, z_n)$ be any two solutions of the equation such that $y_k(t_k) = z_k(t_k)$, $k = 1, 2, \ldots, n$, where the t_k are any n points of the interval $[a,b]$. Then $b - a > 1 \Big/ \sum_{k=1}^{n} l_k$. Hence if

$$f_i(0,t) = 0$$

no nonidentically vanishing solution of the equation can have its components vanishing, respectively, at points inside an interval of length $b - a$ if $b - a$ satisfies the above inequality.   (Fite.)

9. Use the nonlinear integral equation

$$u = \exp\left[-\int_0^t (t - t_1)^2 u(t_1)\, dt_1 \right], \qquad u = w''$$

to establish the existence of the solution of $w''' + 2ww'' = 0$ determined by $w(0) = w'(0)$, $w''(0) = 1$, for all $t \geq 0$.   (Weyl.)

BIBLIOGRAPHY

For recent results on the existence and uniqueness of solutions of first-order systems, we refer to:

La Salle, J. P., *Uniqueness theorems and successive approximations*, Ann. of Math., vol. 50 (1949), pp. 722–730.   References to previous papers will be found there.

The reader will also find it profitable to consult:

Ince, E. L., *Ordinary differential equations*, London, 1927 (Dover reprint, 1944).

Kamke, E., *Differentialgleichungen, Lösungmethoden und Lösungen*, Leipzig, 1943.

Sansone, G., *Equazioni differenziale nel campo reala*, Bologna, 1949.

CHAPTER 4

THE STABILITY OF SOLUTIONS
OF NONLINEAR DIFFERENTIAL EQUATIONS

1. Introduction. In this chapter we begin the study of the stability of the solutions of nonlinear differential equations. We shall consider only systems of the form

$$(1) \qquad \frac{dz_i}{dt} = \sum_{j=1}^{n} a_{ij}(t)z_j + f_i(z,t), \qquad i = 1, 2, \ldots, n$$

where the $f_i(z,t)$ are nonlinear functions of the z_j. The most important case is that where the f_i are independent of t and are power series in the components of z with no zero- or first-order terms. Employing vector-matrix notation, we may write (1) as

$$(2) \qquad \frac{dz}{dt} = A(t)z + f(z)$$

The nonlinearity condition we shall impose is

$$(3) \qquad \frac{\|f(z)\|}{\|z\|} \to 0 \text{ as } \|z\| \to 0$$

If we wish to present results of any comprehensiveness, we are forced to particularize still further and demand that either

(4) (a) $A(t)$ be constant, or
  (b) $A(t)$ be periodic, or
  (c) $A(t)$ be asymptotic to a matrix of either of the above two types

An important category not included above is that of almost-periodic matrices. Despite the obvious importance of equations of this type and the attention that has been paid to the problem, the results are still incomplete. Consequently, we shall not discuss any of the known results here.

From the nonlinearity property of $f(z)$, as expressed by (3), we see that $z = 0$ is a solution of (2). We shall call this the *null*, or *trivial*, solution and shall be interested in solutions of (2) which remain close to $z = 0$. This intuitive notion of closeness will be made precise below.

Our discussion of nonlinear equations will be notably incomplete in its lack of reference to periodic solutions. Despite the apparent simplicity of the concept, the theory of periodic solutions is one of the most difficult of present-day analytical theories and relies upon many deep and complicated theorems of topology. Consequently, we feel that its treatment would be out of place in the introductory volume.

2. Stability. Let us begin by giving a definition of the way we shall employ the overworked word "stability."

Definition. *A solution* $z = (z_1, z_2, \ldots, z_n)$ *of* (2) *of Sec. 1 is said to be stable if for every* $\epsilon > 0$, *there is a* $\delta = \delta(\epsilon)$ *such that any other solution of* (2), $y = (y_1, y_2, \ldots, y_n)$, *for which* $\|z - y\| \leq \delta$ *at* $t = t_0$, *satisfies the further inequality* $\|z - y\| \leq \epsilon$ *for* $t \geq t_0$.

Geometrically put, one thinks of the solution z as a curve in n-dimensional space surrounded by a tubing which has the property that any solution which once penetrates this tubing must thereafter remain within a slightly larger tubing.

Let us now show how the question of the stability of any solution can always be made to depend upon the stability of the null solution $w = 0$ of a related equation. Let z be a solution of $dz/dt = f(z)$ whose stability is to be investigated. Set $y = z + w$, where y is another solution of the equation. Then

$$(1) \qquad \frac{dy}{dt} = \frac{dz}{dt} + \frac{dw}{dt} = f(z + w) = f(z) + J(f,z)w + \cdots$$

where $J(f,z)$ is the Jacobian matrix of f with respect to z. The resulting equation for w is

$$(2) \qquad \frac{dw}{dt} = J(f,z)w + \cdots$$

an equation of the type of (2) of Sec. 1.

It is reasonable to suspect that the stability of the null solution $w = 0$ of (2) is strongly dependent upon the stability of the null solution of the linear approximation

$$(3) \qquad \frac{dw}{dt} = J(f,z)w$$

We shall demonstrate, with the aid of some additional assumptions concerning the linear equation, that this dependence is almost equivalence in the cases where J is a constant or periodic matrix. And then we shall show, by means of a counterexample, that the natural conjecture that the stability of the solution $w = 0$ of the linear equation (3) always implies the same for (2) is not true.

In many interesting cases, a change of variable will transform an equation of general type into one of the special types which may be handled by the methods we present here. In a later chapter, we shall have examples of this when discussing

$$(4) \qquad \frac{d^2u}{dt^2} - a(t)u = 0$$

and

$$(5) \qquad \frac{d}{dt}\left(k(t)\,\frac{du}{dt}\right) + l(t)u^n = 0$$

3. A Preliminary Result. A further motivation for the study of the stability of trivial solutions is furnished by the following result:

Lemma 1. *Consider the system*

$$(1) \qquad \frac{dz_i}{dt} = f_i(z_1, z_2, \ldots, z_n), \qquad i = 1, 2, \ldots, n$$

each f_i being independent of t and continuous in the z_i for $-\infty < z_i < \infty$. If $z = (z_1, z_2, \ldots, z_n)$ is a solution of equation (1) which approaches a constant vector $c = (c_1, c_2, \ldots, c_n)$ as $t \to \infty$, then

$$(2) \qquad f_i(c_1, c_2, \ldots, c_n) = 0, \qquad i = 1, 2, \ldots, n$$

Proof. There are two possibilities for each component z_i. Either it approaches c_i monotonically as $t \to \infty$, or it oscillates infinitely often about c_i. In the first case $dz_i/dt \to 0$ as $t \to \infty$; in the second case $dz_i/dt = 0$ infinitely often. In both cases, it follows that

$$f_i(c_1, c_2, \ldots, c_n) = 0$$

The change of variable $z_i = c_i + w_i$ converts (1) into an equation which has the null vector $w = 0$ as a solution.

Suppose now that we have a mechanical system S, specified by n parameters $z_1, z_2, \ldots, z_n$, whose behavior as functions of time is determined by a set of equations of the type of (1). If we consider an equilibrium position to exist when the z_i are constants independent of time, equation (2) furnishes all possible such states. The question now arises as to what happens to the system when we disturb it slightly, that is, when we change the parameters $c_1, c_2, \ldots, c_n$ into a nearby set of values $c_1', c_2', \ldots, c_n'$.

The following possibilities exist:

1. As $t \to \infty$, the solution of (1) subject to the initial conditions $z_i(0) = c_i'$, $i = 1, 2, \ldots, n$, approaches the stationary solution $z_i = c_i$, the original equilibrium state.

2. As $t \to \infty$, the solution approaches another equilibrium state.

3. As $t \to \infty$, the solution approaches no equilibrium state in the above sense. It may approach a periodic solution or have a more complicated type of behavior.

Let us give some simple examples of the above. Consider the equation

$$(3) \qquad \frac{du}{dt} + u = 0$$

A stationary solution is $u = 0$. If this solution is perturbed, giving rise to another solution with an initial condition $u(t_0) = c_1 \neq 0$ at some time t_0, as $t \to \infty$, this new solution will approach the stationary solution. In this case we would say that the system is in stable equilibrium.

Consider the equation

$$(4) \qquad \frac{du}{dt} = u - u^2$$

There are two equilibrium states, $u = 0$ and $u = 1$ (see Fig. 1). If $u(0) > 0$, $\lim u = 1$ as $t \to \infty$. If $u(0) < 0$, $u \to -\infty$ as $t \to \infty$. The state $u = 0$ is an unstable state, while the state $u = 1$ is stable, with regard to small perturbations.

Finally consider the equation

$$(5) \qquad \frac{d^2u}{dt^2} + u = 0$$

The solution $u = 0$ is a stationary one. In one sense it is stable; in another sense it is unstable. If we consider the solution

$$u = c_1 \cos t + c_2 \sin t$$

corresponding to $u(0) = c_1$, $u'(0) = c_2$, u will not approach zero as $t \to \infty$. However, u will remain as close as desired to the stationary solution if we take $|c_1| + |c_2|$ sufficiently small. According to our definition of stability given in Sec. 2, the stationary solution is stable.

4. Fundamental Stability Theorem, First Proof. In this section we present the fundamental result connecting the stability of the null solution of the nonlinear equation with the behavior of the solutions of the linear equation.

We write our equation in the form

$$(1) \qquad \frac{dz}{dt} = Az + f(z), \qquad z(0) = c$$

where A is a constant matrix and $f(z)$ satisfies the nonlinearity condition $\|f(z)\|/\|z\| \to 0$ as $\|z\| \to 0$.

Heuristically, we might argue as follows: If $\|z(0)\|$ is small, then, by virtue of the nonlinearity condition, $Az + f(z)$ is very nearly Az. If all the solutions of $dy/dt = Ay$ approach zero as $t \to \infty$, z should have no opportunity ever to become large. Hence, for all t, z should act like a solution of $dy/dt = Ay$.

This "proof" happens to be correct if A is a constant matrix. As we shall see by means of a specific example, it is not true in general if the coefficient matrix is variable.

Let us now prove

Theorem 1. *If*

(2) (a) *Every solution of $dy/dt = Ay$ approaches zero as $t \to \infty$*
     (b) *$f(z)$ is continuous in some region about $z = 0$*
     (c) *$\|f(z)\|/\|z\| \to 0$ as $\|z\| \to 0$*

then $z = 0$ is a stable solution of (1).

Furthermore, every solution of (1) *for which $\|z(0)\|$ is sufficiently small approaches zero as $t \to \infty$.*

Three proofs of this fundamental theorem will be given in this chapter. Each proof has its own special interest and its own range of generalization.

Note that nothing is said concerning the uniqueness of solution. In general, we may expect nonuniqueness, and the interesting point is that the theorem asserts that no continuation can deviate too far from the trivial solution if the initial value is close enough to the origin.

First Proof. It is a consequence of our fundamental existence theorem that (1) possesses a solution in some neighborhood of $t = 0$, say $0 \leq t \leq t_0$. We wish to show that any such solution can be continued over the entire positive t interval. To do this, it is sufficient to show that $\|z\|$ is uniformly bounded for $t \geq 0$, and that the bound thus obtained lies within the region where $f(z)$ is continuous.

The first part of our hypothesis tells us that the solution of

$$(3) \qquad \frac{dY}{dt} = AY, \qquad Y(0) = I$$

approaches zero as $t \to \infty$, which, in turn, implies that

$$(4) \qquad \int^{\infty} \|Y(t)\| \, dt < \infty$$

Since the solution of

$$(5) \qquad \frac{dy}{dt} = Ay, \qquad y(0) = c$$

is given by $y = Yc$, we see that $\|y\| \leq \|Y\|\|c\| \leq a_1\|c\|$.

Applying Theorem 4 of Chap. 1, we conclude that the nonlinear differential equation (1) may be converted into the nonlinear integral equation

$$(6) \qquad z = y + \int_0^t Y(t - t_1) f(z(t_1)) \, dt_1$$

We now derive a uniform bound for any solution of (6), namely, $\|z\| < 2a_1\|c\|$, provided that $\|c\|$ is sufficiently small. This will prove the desired stability. The proof is by contradiction. Let t_2 be the first point at which $\|z\| = 2a_1\|c\|$ in the interval $(0, t_0)$. That $t_2 > 0$ follows from the condition $z(0) = y(0)$. At the point t_2, we have

$$(7) \qquad 2a_1\|c\| = \|z\| \leq \|y\| + \int_0^{t_2} \|Y(t_2 - t_1)\| \|f(z(t_1))\| \, dt_1$$

If $\|c\|$ is small enough, we have, by virtue of the second part of our hypothesis, $\|f(z(t))\| \leq \epsilon_1 \|z(t)\|$ for $0 \leq t \leq t_2$, where ϵ_1 can be made as small as desired by suitable choice of $\|c\|$. Thus

$$(8) \qquad 2a_1\|c\| \leq a_1\|c\| + \epsilon_1(2a_1\|c\|) \int_0^{t_2} \|Y(t_2 - t_1)\| \, dt_1$$
$$\leq a_1\|c\| + \epsilon_1(2a_1\|c\|) \int_0^\infty \|Y(t_1)\| \, dt_1 < 2a_1\|c\|$$

if $\|c\|$ is sufficiently small. Consequently there is no point t_2. The solution may now be continued, interval by interval, preserving the uniform bound, until we have covered the positive t axis.

To show that $\|z\| \to 0$ as $t \to \infty$ if $\|z(0)\|$ is sufficiently small, we make the change of variable, $z = xe^{\lambda t}$, where λ is a fixed quantity less that zero and greater than the real part of any characteristic root of A. Once we fix this value of λ, the upper bound on $\|x\|$ will depend upon λ. The new variable x satisfies the equation

$$(9) \qquad \frac{dx}{dt} = (A - \lambda I)x + e^{-\lambda t}f(xe^{\lambda t})$$

Although the form of the nonlinear term is slightly different from that treated before, there is no trouble in verifying that the same argument suffices, because of the assumption (2b). Since x is uniformly bounded, it follows that $\|z\| \to 0$.

Exercise

Show that the condition that $f(z)$ be continuous may be relaxed considerably.

---

5. Fundamental Stability Theorem, Second Proof. Our principal tool, which will be used again subsequently, is the matrix-transformation theorem of Sec. 8 of Chap. 1, according to which we may find a constant matrix T having the property that

$$(1) \qquad T^{-1}AT = \begin{pmatrix} \lambda_1 & b_{12} & \cdots & b_{1n} \\ 0 & \lambda_2 & \cdots & b_{2n} \\ \cdot & & & \cdot \\ \cdot & & & \cdot \\ \cdot & & & \cdot \\ 0 & 0 & \cdots & \lambda_n \end{pmatrix}$$

where the elements of the main diagonal are the characteristic roots of A, and where $|b_{ij}| \le \epsilon$, where ϵ is any prescribed positive quantity.

The substitution $z = Tx$ yields

$$(2) \qquad \frac{dx}{dt} = T^{-1}ATx + T^{-1}f(Tx)$$

Let $f_1(x) = T^{-1}f(Tx)$. It is easily seen that $f_1(x)$ satisfies the same condition as $f(x)$. Turning to the individual components of x, we have

$$(3) \qquad \frac{dx_1}{dt} = \lambda_1 x_1 + \sum_{j>1} b_{1j}x_j + f_{11}(x)$$

$$\frac{dx_2}{dt} = \lambda_2 x_2 + \sum_{j>2} b_{2j}x_j + f_{12}(x)$$

$$\frac{dx_n}{dt} = \lambda_n x_n + f_{1n}(x)$$

Consider the sum $\sum_{k=1}^{n} |x_k|^2$. Since

$$(4) \quad \frac{d}{dt}|x_k|^2 = \bar{x}_k \frac{dx_k}{dt} + x_k \frac{d\bar{x}_k}{dt}$$

(where by $\bar{x}_k$ we mean the complex conjugate of x_k),

$$= \lambda_k |x_k|^2 + \sum_{j>k} b_{kj}x_j\bar{x}_k + \bar{x}_k f_{1k}(x)$$
$$+ \bar{\lambda}_k |x_k|^2 + \sum_{j>k} \bar{b}_{kj}\bar{x}_j x_k + x_k \overline{f_{1k}(x)}$$

we obtain

$$(5) \quad \frac{1}{2}\frac{d}{dt}\left(\sum_{k=1}^{n}|x_k|^2\right) \le \text{Re }(\lambda)\left(\sum_{k=1}^{n}|x_k|^2\right) + \epsilon\|x\|^2 + \|x\|\|f_1(x)\|$$

where λ is the characteristic root with least negative real part.

At $t = 0$, we have, since Re $(\lambda) < 0$, and $\|x(0)\|$ is sufficiently small, that

$$(6) \qquad \text{Re } (\lambda) \left(\sum_{k=1}^{n} |x_k|^2 \right) + \epsilon \|x\|^2 + \|x\| \|f_1(x)\| < 0$$

Thus $\sum_{k=1}^{n} |x_k|^2$ is decreasing in the immediate neighborhood of $t = 0$, and the argument may now be repeated, so as to obtain the following inequality for all t:

$$(7) \qquad \sum_{k=1}^{n} |x_k|^2 \leq \left(\sum_{k=1}^{n} |x_k(0)|^2 \right) e^{-at}$$

for some $a > 0$.

Exercise

Does this proof require that $f(z)$ be continuous?

---

6. Fundamental Stability Theorem, Third Proof. Let us now present a third proof depending upon the method of successive approximations. Since our hypothesis and conclusions are now slightly different, we state the results in detail.

Theorem 1′. *If*

(1) (a) *Every solution of $dy/dt = Ay$ approaches zero as $t \to \infty$*
     (b) $\|f(z)\|/\|z\| \to 0$ *as* $\|z\| \to 0$
     (c) $\|f(z_1) - f(z_2)\| \leq c_1\|z_1 - z_2\|$ *for* $\|z_1\|$ *and* $\|z_2\|$ *less than* c_2, *where* $c_1 \to 0$ *as* $c_2 \to 0$

then $z = 0$ is a stable solution of $dz/dt = Az + f(z)$.

Every solution z for which $\|z(0)\|$ is sufficiently small may be computed by the following method of successive approximations:

$$(2) \qquad \frac{dz_0}{dt} = Az_0, \qquad z_0(0) = c$$

$$\frac{dz_{n+1}}{dt} = Az_{n+1} + f(z_n), \qquad z_{n+1}(0) = c, \qquad n = 0, 1, \ldots$$

and approaches zero as $t \to \infty$.

Proof. We start from the integral equation

$$(3) \qquad z = y + \int_0 Y(t - t_1)f(z(t_1)) \, dt_1$$

and apply the method of successive approximations, as above in (2).

There are two steps to the proof of convergence, which we shall show to be uniform in t over the infinite t interval $0 \leq t \leq \infty$. The first step is to show the uniform boundedness of $\|z_n\|$ over $0 \leq t \leq \infty$, provided as always that $\|z(0)\|$ is sufficiently small. The second step is to establish the uniform convergence of $\sum_{n=0}^{\infty} (z_{n+1} - z_n)$.

We have $\|z_0\| = \|y\| = \|Yc\| \leq \|Y\|\|c\| \leq a_1\|c\| < 2a_1\|c\|$, where $c = z(0) = y(0)$. Let us now show that $\|z_n\| < 2a_1\|c\|$ implies that $\|z_{n+1}\| < 2a_1\|c\|$. From (2),

$$
(4) \qquad
\begin{aligned}
\|z_{n+1}\| &\leq \|y\| + \int_0^t \|Y(t - t_1)\|\|f(z_n)\|\, dt_1 \\
&\leq a_1\|c\| + \epsilon_1 \int_0^t \|Y(t - t_1)\|\|z_n\|\, dt_1
\end{aligned}
$$

provided that $\|c\|$ is small enough, and thus

$$
(5) \qquad
\begin{aligned}
\|z_{n+1}\| &\leq a_1\|c\| + \epsilon_1(2a_1\|c\|) \int_0^t \|Y(t - t_1)\|\, dt_1 \\
&\leq a_1\|c\| + \epsilon_1(2a_1\|c\|) \int_0^{\infty} \|Y(t_1)\|\, dt_1 < 2a_1\|c\|
\end{aligned}
$$

if $\|c\|$, and consequently ϵ_1, are small enough. Since the inequality holds for $n = 0$, it holds for all $n > 0$.

Now let us show the convergence of $\sum_{n=0}^{\infty} (z_{n+1} - z_n)$. We have

$$
(6) \qquad
\begin{aligned}
z_{n+1} - z_n &= \int_0^t Y(t - t_1)[f(z_n) - f(z_{n-1})]\, dt_1 \\
\|z_{n+1} - z_n\| &\leq \int_0^t \|Y(t - t_1)\|\|f(z_n) - f(z_{n-1})\|\, dt \\
&\leq c_1 \int_0^t \|Y(t - t_1)\|\|z_n - z_{n-1}\|\, dt
\end{aligned}
$$

using (1c). The constant c_1 can be made arbitrarily small by taking the norm of $\|c\|$, which determines the norm of $\|z\|$, sufficiently small. From (6) we have

$$
(7) \qquad \|z_{n+1} - z_n\| \leq c_1 \Big(\max_{0 \leq t_1 \leq t} \|z_n - z_{n-1}\|\Big) \int_0^t \|Y(t - t_1)\|\, dt_1
$$

and thus,

$$
(8) \qquad \max_{0 \leq t \leq t_1} \|z_{n+1} - z_n\| \leq \Big(c_1 \int_0^{\infty} \|Y(t_1)\|\, dt_1\Big) \max_{0 \leq t \leq t_1} \|z_n - z_{n-1}\|
$$

Since $c_1 \int_0^{\infty} \|Y(t_1)\|\, dt_1 = c_3$ will be less than 1 for c_2 chosen small enough, the series

$$(9) \qquad \sum_{n=0}^{\infty} \max_{0 \leq t \leq t_1} \|z_{n+1} - z_n\|$$

converges by comparison with the series $\sum_{n=1}^{\infty} c_3^n$. This convergence is uniform over the positive half of the real axis. Thus z_n converges uniformly over this interval to a limit function z, which by virtue of (2) satisfies the equation

$$(10) \qquad z = y + \int_0^t Y(t - t_1) f(z) \, dt_1$$

and thus satisfies our original differential equation.

Since $\|f(z)\| \leq \epsilon_1 \|z\|$, we obtain

$$(11) \qquad \begin{aligned} \|z\| &\leq \|y\| + \epsilon_1 \int_0^t \|Y(t - t_1)\| \|z\| \, dt_1 \\ &\leq c_4 e^{-at} + \epsilon_1 \int_0^t e^{-a(t-t_1)} \|z\| \, dt_1 \end{aligned}$$

for some positive constants c_4 and $a > 0$. Hence

$$(12) \qquad \|z\| e^{at} \leq c_4 + \epsilon_1 \int_0^t e^{at_1} \|z\| \, dt_1$$

whence the fundamental lemma yields $\|z\| e^{at} \leq c_4 e^{\epsilon_1 t}$. If ϵ_1 is sufficiently small, we see that $\|z\| \to 0$ as $t \to \infty$.

Exercises

1. Show that the above result holds for the equation

$$\frac{dz}{dt} = (A + B(t))z + f(z)$$

provided that $\|B(t)\|$ is sufficiently small, or that $\int^{\infty} \|B\| \, dt < \infty$. Show how this result may be used to reduce problems of this type to the case where A has distinct characteristic roots.

2. Show that the above result holds for the equation

$$\frac{dz}{dt} = Az + f\left(z, \frac{dz}{dt}\right)$$

provided that $f(z, dz/dt)$ satisfies suitable conditions.

3. If $|u(0)|$ is sufficiently small, is the solution of $du/dt = -2u + e^t u^2$ bounded? Generalize.

4. If $u' + u = u''^2$, is it true that, if $|u(0)|$ is sufficiently small and if $u'(0)$ is chosen properly, there is a solution which approaches zero as $t \to \infty$?

5. Consider the nonlinear system

$$\frac{dy_i}{dt} + y_i \sum_{j=1}^{n} a_{ij} y_j = \sum_{j,k=1}^{n} b_{ijk} y_j y_k, \qquad i = 1, 2, \ldots, n$$

where $b_{ijk} > 0$, and where the above equations imply a relation of the form $\sum_{i=1}^{n} a_i y_i = c_1$. Show that under these assumptions every solution for which $y_i(0) \geq 0$ may be continued throughout $0 \leq t < \infty$ and the solutions will remain nonnegative and uniformly bounded. (Carleman.)

---

7. Asymptotic Behavior of the Solutions. We know that the solutions of

$$(1) \qquad \frac{dz}{dt} = Az + f(z)$$

under the hypotheses of Theorem 1 or 1′ approach zero and are actually majorized by exponentials of the form e^{-at}, with $a > 0$. The question arises as to the precise asymptotic behavior of the solutions.

Let us consider the most important case where the hypotheses are satisfied, the case where the components of $f(z)$ are power series, in the components of z, lacking constant or first-degree terms.

We may then write (1) in the form

$$(2) \qquad \frac{dz}{dt} = (A + G(z))z$$

where $G(z)$ is a matrix whose elements are power series, in the components of z, lacking constant terms. Since $\|z\| \to 0$ as $t \to \infty$, (2) has the form

$$(3) \qquad \frac{dz}{dt} = (A + B(t))z$$

where $B(t) \to 0$ as $t \to \infty$, and where actually $\|B(t)\| \leq c_1 e^{-at}$, $a > 0$, for $t \geq 0$, together with similar conditions on the derivative.

If A has simple characteristic roots, the asymptotic behavior of the solutions of (3) may be immediately deduced from the results of Chap. 2. If A has multiple roots, the same techniques, aided by the strong bounds on $B(t)$ and $B'(t)$, will readily yield the asymptotic behavior. We leave arrival at the precise results as exercises for the reader.

8. Periodic Coefficients. We have seen in Chap. 2 that the analogues of the boundedness, stability, and asymptotic-behavior theorems for linear differential equations with almost-constant coefficients flow easily from the representation theorem for solutions of linear equations with periodic coefficients; in the same way we shall show now, using the same weapon, that the analogue of Theorem 1 holds in the case where $A(t)$ is periodic.

Theorem 2.   *Consider the equation*

$$(1) \qquad \frac{dz}{dt} = A(t)z + f(z)$$

where $A(t)$ is a periodic matrix of period τ.

The trivial solution $z = 0$ is stable provided that

(2)   (a) *Every solution of $dy/dt = A(t)y$ approaches zero as $t \to \infty$*
       (b) *$f(z)$ is continuous in some region about $z = 0$*
       (c) *$\|f(z)\|/\|z\| \to 0$ as $\|z\| \to 0$*

Proof.   Let Y be the solution of

$$(3) \qquad \frac{dY}{dt} = A(t)Y, \qquad Y(0) = I$$

and let y be the solution of

$$(4) \qquad \frac{dy}{dt} = A(t)y, \qquad y(0) = z(0)$$

Then z satisfies the familiar integral equation

$$(5) \qquad z = y + \int_0^t Y(t)Y^{-1}(t_1)f(z)\, dt_1$$

Employing the representation theorem $Y = P(t)e^{Bt}$, (5) becomes

$$(6) \qquad z = y + \int_0^t P(t)e^{B(t-t_1)}P(t_1)^{-1}f(z)\, dt_1$$

The hypotheses of (2) imply that $\|e^{Bt}\| \to 0$ as $t \to \infty$, and that

$$(7) \qquad \int^\infty \|e^{Bt_1}\|\, dt_1 < \infty$$

From (6) we derive

$$(8) \qquad \|z\| \le \|y\| + c_1^2 \int_0^t \|e^{B(t-t_1)}\|\,\|f(z)\|\, dt_1$$

and the proof now proceeds as in Sec. 4.

The device used in that section permits us to conclude that $z \to 0$ as $t \to \infty$.

Exercise

Carry through the proof using the technique of Sec. 5.

---

9. Counterexample to a Proposed General Stability Theorem. It might be expected on the basis of previous results that for stability of the trivial solution $z = 0$ of a system of nonlinear equations, it would be sufficient to have the solutions of the linear approximation possess the property of tending to zero as $t \to \infty$. We shall show by an example that no such general result can hold.

The solution of

$$(1) \qquad \frac{dy_1}{dt} = -ay_1, \qquad\qquad y_1(0) = c_1$$

$$\frac{dy_2}{dt} = [(\sin \log t + \cos \log t) - 2a]y_2, \qquad y_2(0) = c_2$$

is

$$(2) \qquad\qquad y_1 = c_1 e^{-at}$$
$$y_2 = c_2 e^{(t \sin \log t - 2at)}$$

which approaches zero as $t \to \infty$ for $a > \frac{1}{2}$.

On the other hand the solution of the nonlinear system

$$(3) \qquad \frac{dz_1}{dt} = -az_1, \qquad\qquad z_1(0) = c_1$$

$$\frac{dz_2}{dt} = (\sin \log t + \cos \log t - 2a)z_2 + z_1^2, \qquad z_2(0) = c_2$$

is

$$(4) \qquad\qquad z_1 = c_1 e^{-at}$$
$$z_2 = e^{t \sin \log t - 2at} \left(c_2 + c_1^2 \int_0^t e^{-t_1 \sin \log t_1} \, dt_1 \right)$$

which, we we shall show, approaches zero as $t \to \infty$ only if $c_1 = 0$, provided that we choose $1 + e^{-\pi/2} > 2a > 1$. Therefore, choosing $\|z(0)\|$ sufficiently small does not suffice to have the solution of the nonlinear equation also approach zero. Furthermore, we shall show that the solution is actually unbounded as $t \to \infty$. We have

$$(5) \qquad \int_0^t e^{-t_1 \sin \log t_1} \, dt_1 > \int_{te^{-\pi}}^{te^{-2\pi/3}} e^{-t_1 \sin \log t_1} \, dt_1 > \exp\, (te^{-\pi/2}) \int_{te^{-\pi}}^{te^{-2\pi/3}} dt_1$$
$$= t(e^{-2\pi/3} - e^{-\pi}) \exp\, (te^{-\pi/2})$$

for $t = e^{(2n+\frac{1}{2})\pi}$. Thus at this point,

$$(6) \qquad e^{t \sin \log -2at} \int_0^t e^{-t_1 \sin \log t_1} \, dt > c_3 t \exp\, [(1 + e^{-\pi/2} - 2a)t]$$

Consequently, $z_2 \to 0$ as $t \to \infty$ only if $c_1 = 0$.

10. Instability.   We have seen in the previous sections that $z = 0$ is a stable solution of

$$(1) \qquad \frac{dz}{dt} = Az + f(z)$$

provided that we have imposed the proper conditions upon A and $f(z)$.

Let us show by a simple example that some restriction on the magnitude of $\|z(0)\|$ is necessary to ensure the boundedness of the solution, even when the characteristic roots of A have negative real parts and when the nonlinear terms satisfy the usual conditions.   Consider

$$(2) \qquad \frac{du}{dt} = -u + u^2, \qquad u(0) = a > 1$$

whose solution is given by

$$(3) \qquad u = \frac{a}{a - (a - 1)e^t}$$

As $t \to \log (a/a - 1)$, u evidently becomes unbounded.

Conversely, a positive characteristic root does not necessitate unbounded solutions.   For example, if

$$(4) \qquad \frac{du}{dt} = u - u^2, \qquad 0 < u(0) < \infty$$

we see that $u \to 1$ as $t \to \infty$.

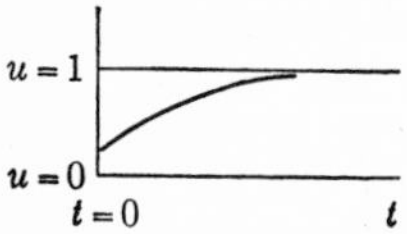

We can state the following result:

Theorem 3.   *If*

Fig. 1.

(5)   (a) *A possesses at least one characteristic root with positive real part*

      (b) $\|f(z)\|/\|z\| \to 0$ *as* $\|z\| \to 0$

then $z = 0$ is unstable.

Proof.   We use a method of Sec. 5.   Let T be a transformation such that

$$(6) \qquad T^{-1}AT = \begin{pmatrix} \lambda_1 & b_{12} & \cdots & b_{1n} \\ 0 & \lambda_2 & \cdots & b_{2n} \\ \cdot & \cdot & & \cdot \\ \cdot & \cdot & & \cdot \\ \cdot & \cdot & & \cdot \\ 0 & 0 & \cdots & \lambda_n \end{pmatrix}$$

where $|b_{ij}| \leq \epsilon$, ϵ being a positive quantity to be prescribed.   Let $z = Tx$. Then

$$(7) \qquad \frac{dx_1}{dt} = \lambda_1 x_1 + b_{12}x_2 + \cdots + b_{1n}x_n + g_1(x)$$

$$\frac{dx_2}{dt} = \qquad\quad \lambda_2 x_2 + \cdots + b_{2n}x_n + g_2(x)$$

$$\cdot$$
$$\cdot$$
$$\cdot$$

$$\frac{dx_n}{dt} = \qquad\qquad\qquad\qquad\quad \lambda_n x_n + g_n(x)$$

Since the order in which we take the characteristic roots is immaterial, let us assume that Re $(\lambda_1) > 0, \ldots,$ Re $(\lambda_k) > 0$, where $k \geq 1$. Multiplying the kth equation by $\bar{x}_k$ if λ_k has a positive real part, and by $-\bar{x}_k$ if it has a nonpositive real part, we obtain, using the condition

$$\frac{\|f(z)\|}{\|z\|} \to 0$$

as $\|z\| \to 0$,

$$(8) \quad \frac{d}{dt}\left(|x_1|^2 + |x_2|^2 + \cdots + |x_k|^2 - |x_{k+1}|^2 - \cdots - |x_n|^2\right)$$
$$= \text{Re }(\lambda_1)|x_1|^2 + \text{Re }(\lambda_2)|x_2|^2 + \cdots + \text{Re }(\lambda_k)|x_k|^2$$
$$- \text{Re }(\lambda_{k+1})|x_{k+1}|^2 - \cdots - \text{Re }(\lambda_n)|x_n|^2 + \cdots$$

where the additional terms are of smaller order of magnitude.

Again we argue by contradiction. If the trivial solution is stable, by choosing $\|x(0)\|$ sufficiently small we can ensure $\|x\| \leq \epsilon$ for $t \geq 0$. From (8) we then conclude that

$$(9) \quad \frac{d}{dt}\left(|x_1|^2 + \cdots + |x_k|^2 - |x_{k+1}|^2 - \cdots - |x_n|^2\right)$$
$$\geq c_1(|x_1|^2 + |x_2|^2 + \cdots + |x_k|^2)$$
$$\geq c_1(|x_1|^2 + \cdots + |x_k|^2 - |x_{k+1}|^2 - \cdots - |x_n|^2)$$

with $c_1 > 0$, and thus

$$(10) \quad |x_1|^2 + \cdots + |x_k|^2 - |x_{k+1}|^2 - \cdots - |x_n|^2$$
$$\geq (|x_1(0)|^2 + \cdots + |x_k(0)|^2 - |x_{k+1}(0)|^2 - \cdots - |x_n(0)|^2)e^{c_1 t}$$

If we now choose

$$|x_1(0)|^2 + \cdots + |x_k(0)|^2 - |x_{k+1}(0)|^2 - \cdots - |x_n(0)|^2 > 0$$

we see that (10) contradicts the inequality $\|x\| \leq \epsilon$ for t sufficiently large.

11. Conditional Stability. Despite the negative character of the result of the previous section, we shall show that if there are any characteristic roots of A with negative real parts, then the solution $z = 0$ possesses a certain conditional stability.

   Theorem 4. *If*

(1)   (a) *k of the characteristic roots of A have negative real parts, where $k \leq n$*
       (b) *$\|f(z)\|/\|z\| \to 0$ as $\|z\| \to 0$*
       (c) *$\|f(z_1) - f(z_2)\| \leq c_1\|z_1 - z_2\|$ for $\|z_1\|$ and $\|z_2\| \leq c_2$, where $c_1 \to 0$ as $c_2 \to 0$*

there is a k-parameter family of solutions of

$$(2) \qquad \frac{dz}{dt} = Az + f(z)$$

which approach zero as $t \to \infty$.

   Proof. Let y and Y have their previous connotation and consider the integral equation

$$(3) \qquad z = y + \int_0^t Y(t - t_1)f(z)\,dt_1$$

   To obtain solutions of (3) which approach zero as $t \to \infty$, we must somehow weed out the elements of $Y(t)$ which do not tend to zero as $t \to \infty$. To this end we consider the following decomposition:

$$(4) \qquad Y = Y_1 + Y_2$$

with $Y_1 = (u_{ij})$ and $Y_2 = (v_{ij})$, where $y_{ij} = u_{ij} + v_{ij}$, here u_{ij} is the part of y_{ij} which tends to zero as $t \to \infty$, and v_{ij} is the remaining part, corresponding to the characteristic roots of A which have zero or positive real parts. It is easy to see that

$$(5) \qquad \frac{dY_1}{dt} = AY_1, \qquad \frac{dY_2}{dt} = AY_2$$

   Using (4), (3) may be written

$$
\begin{aligned}
(6) \quad z &= y + \int_0^t Y_1(t - t_1)f(z)\,dt_1 + \int_0^t Y_2(t - t_1)f(z)\,dt_1 \\
&= y + \int_0^t Y_1(t - t_1)f(z)\,dt_1 - \int_t^\infty Y_2(t - t_1)f(z)\,dt_1 \\
&\qquad\qquad\qquad\qquad + \int_0^\infty Y_2(t - t_1)f(z)\,dt_1
\end{aligned}
$$

   At the moment, these operations are purely formal, since we do not know whether the integrals from 0 to ∞ and from t to ∞ exist. If, however, the last integral exists, it is a solution of $dy/dt = Ay$. We may then combine it with the first term to form a new y, and consider the integral equation

$$(7) \qquad z = y + \int_0^t Y_1(t - t_1)f(z)\, dt_1 - \int_t^\infty Y_2(t - t_1)f(z)\, dt_1$$

To obtain a solution of (7), we employ the method of successive approximations. Reversing the operations above, we see that any solution of (7) is a solution of the integral equation (3), with a different y, and consequently is a solution of (2).

Choose the y in (7) to be any element of the k-parameter family of solutions of $dy/dt = Ay$ which tend to zero as $t \to \infty$, and let $\|y(0)\|$ be a small constant whose precise magnitude will be specified subsequently. Since $y \to 0$ as $t \to \infty$, $\|y\| \le c_3 e^{-at}$ for $t \ge 0$ and for some $a > 0$, and c_3 can be made as small as desired by an appropriate choice of $\|y(0)\|$. Furthermore, there exists an $a_1 > a$ such that $\|Y_1(t)\| \le c_4 e^{-a_1 t}$, and finally $\|Y_2(t)\| \le c_5 e^{bt}$ for some constant b. We now proceed by induction. Set

$$(8) \qquad z_0 = y$$

$$z_{n+1} = y + \int_0^t Y_1(t - t_1)f(z_n)\, dt_1 - \int_t^\infty Y_2(t - t_1)f(z_n)\, dt_1$$

From the inequality for y, it follows that $\|z_0\| \le c_3 e^{-at}$. Let us show that $\|z_n\| \le 2c_3 e^{-at}$ for $n \ge 0$. From (8),

$$(9) \qquad \|z_{n+1}\| \le c_3 e^{-at} + 2c_3 c_4 \epsilon \int_0^t e^{-a_1(t-t_1)} e^{-at_1}\, dt_1 + 2c_3 c_5 \epsilon \int_t^\infty e^{b(t-t_1)} e^{-at_1}\, dt_1$$

using the condition (1b). If ϵ is sufficiently small, we obtain $\|z_{n+1}\| \le 2c_3 e^{-at}$.

The uniform convergence of $\displaystyle\sum_{n=0}^\infty \|z_{n+1} - z_n\|$ now follows as in Sec. 6, and the remainder of the proof proceeds similarly.

12. A Particular Case of Zero Characteristic Roots. Any discussion of the case where A has characteristic roots with zero real parts is extremely difficult, and for that reason we shall not go into it here. There are, however, a few special cases where the behavior of the solutions as $t \to \infty$ may be determined. One is the following:

Theorem 5. *Consider the differential equation*

$$(1) \qquad \frac{dz}{dt} = Az + f(z,t)$$

where

(2) (a) *All solutions of $dy/dt = Ay$ are bounded*

     (b) $\|f(z,t)\|/\|z\| \le c_1 g(t)$ *for* $\|z\| \le c_2$, *where* $\displaystyle\int^\infty g(t)\, dt < \infty$

Under these conditions, $z = 0$ is a stable solution.

We leave the proof as an exercise.

13. Difference Equations. It is possible to extend many of the results of this chapter and of previous chapters to difference equations of the form

(1) $$z(t + h) = Az(t) + f(z(t)), \qquad t = 0, h, 2h, \ldots$$

The method used to establish these analogues is abstractly equivalent to that used for differential equations. Consequently, we shall present the various steps as exercises.

Exercises

1. What is a necessary and sufficient condition that all solutions of $y(t + h) = Ay(t)$ approach zero as $t \to \infty$?

2. Find the formula expressing the solutions of $z(t + h) = Az(t) + w(t)$ in terms of the solution of $y(t + h) = Ay(t)$, that is, the analogue of Theorem 3 of Chap. 1.

3. Find the nonlinear sum equation equivalent to (1). Use this equation to prove the analogue of Theorem 1.

4. Use this result to prove Theorem 1, by means of a limiting process.

5. Show that, if $\displaystyle\sum_{n=1}^{\infty} \|B(nh)\| < \infty$ and if all solutions of

$$y(t + h) = Ay(t)$$

are bounded, then all solutions of $z(t + h) = (A + B(t))z(t)$ are bounded.

NOTE: In the above exercises t takes on only the values $t = 0, h, 2h, \ldots$.

BIBLIOGRAPHY

Some recent important books which discuss other mathematical aspects of the theory of linear and nonlinear differential equations and physical applications of the theory are:

Andronow, A. A., and C. E. Chaikin, *Theory of oscillations*, Princeton University Press, Princeton, N.J., 1949.

Kryloff, N., and N. Bogoliuboff, *Introduction to non-linear mechanics*, Princeton University Press, Princeton, N.J., 1943.

Lefschetz, S., *Lectures on differential equations*, Princeton University Press, Princeton, N.J., 1946.

———, *Contributions to the theory of non-linear equations*, Princeton University Press, Princeton, N.J., 1950.

Minorsky, N., *Introduction to non-linear mechanics*, Edwards Bros., Inc., Ann Arbor, Mich., 1947.

Rocard, Y., *Dynamique générale des vibrations*, Masson et Cie, Paris, 1949.

Stoker, J. J., *Non-linear vibrations in mechanical and electrical systems*, Interscience Publishers, Inc., New York, 1950.

Section 4. The theorem of this section is the fundamental result of Liapounoff and Poincaré. The original statements and proofs may be found in:

Liapounoff, A., *Problème général de la stabilité du mouvement*, Ann. Fac. Sci. Univ. Toulouse, vol. 9 (1907), pp. 203–475. Reprinted in the Annals of Mathematics Studies, 1947.

Poincaré, H., *Méthodes nouvelles de la mécanique céleste*, Vols, I, III, Gauthier-Villars & Cie, Paris, 1892.

The first proof in the text is due to N. Levinson, communicated privately to the author; the second proof is due to O. Perron, Math. Zeit., vol. 29 (1929), pp. 129–160, while the idea of the third proof is due to E. Cotton, Ann. École Norm., ser. 3, vol. 28 (1911), pp. 473–521. See also his book.

Cotton, E., *Approximations successives et les équations différentielles*, Mém. Sci. Fasc., vol. XXVIII, Hermann et Cie., Paris, 1928.

For a general discussion and other methods of proof, see:

Bellman, R., *On the boundedness of solutions of non-linear differential and difference equations.*, Trans. Amer. Math. Soc., vol. 62 (1947), pp. 357–386.

Section 9. See the above reference to O. Perron.

Section 11. See the above reference to O. Perron.

Sections 12 and 13. See the above reference to R. Bellman.

CHAPTER 5

THE ASYMPTOTIC BEHAVIOR OF THE SOLUTIONS
OF SOME NONLINEAR EQUATIONS OF THE FIRST ORDER

1. Introduction. In this chapter we shall consider the question of the asymptotic behavior of solutions of polynomial equations of the type

$$(1) \qquad P(t,u,u') = \Sigma a_{lmn} t^l u^m (u')^n = 0$$

with $l,m,n \geq 0$ and $l + m + n \leq N$, with emphasis upon the important particular case

$$(2) \qquad u' = \frac{P(u,t)}{Q(u,t)}$$

where P and Q are polynomials.

In addition, some corresponding results for the equation

$$(3) \qquad u'' = \frac{P(u,t)}{Q(u,t)}$$

will be indicated. Since the results and methods of proof for this second case are very similar to those for (2), but very much more detailed, we shall content ourselves with stating the results and omitting the proofs.

The general problem of existence, continuation, and analytic nature of the solutions of (1) and (2) is one to which the theory of functions of a complex variable has been applied with some success. Nevertheless, there exists at the present time no general theory of the real solutions of real differential equations, and the problem remains one of great difficulty. To avoid the inherent difficulties, we make a simplifying restriction which enables us to treat many interesting and important cases. We shall study only those solutions of (1) which exist for all sufficiently large t. This focusing of the spotlight of our attention results in a compensating illumination of the subject.

2. Upper Bounds for the Solutions of $P(t,u,u') = 0$. We shall call a solution of (1) of Sec. 1 a *proper* solution if it exists and if it has a continuous derivative for $t \geq t_0$. Henceforth we shall confine ourselves to the study of proper solutions and so, for convenience, shall occasionally drop the adjective.

We begin with the question of upper bounds and demonstrate

Theorem 1. *If $u(t)$ is a proper solution of $P(t,u,u') = 0$, there exists a constant k for which*

$$(1) \qquad |u| \le \exp \frac{t^{k+1}}{k+1}, \qquad t \ge t_1 \ge t_0$$

If m is the exponent of the highest power of t appearing in $P(t,u,u')$, then k may be chosen to be $m + \epsilon$ for any $\epsilon > 0$.

Proof. The proof will be by contradiction. If the result is false, one of the following must occur:

(2) (a) $|u| = \exp (t^{k+1}/k + 1)$ has roots as large as desired
       (b) For some $t_1 \ge t_0$, $|u| > \exp (t^{k+1}/k + 1)$ for $t \ge t_1$

Let us first show that case $(2a)$ leads to a contradiction. Let t_1, t_2, $\ldots$, t_n, $\ldots$, taken in increasing order, with $t_n \to \infty$, be the roots of $u(t) = w(t)$, where we shall set $w(t) \equiv \exp (t^{k+1}/k + 1)$. It is sufficient to consider this case, since the case where $-u(t) = w$ has infinitely many roots may be transformed into the first by the transformation $u \to -u$.

Assume first that there are an infinity of consecutive roots occurring as consequences of intersections of $u(t)$ and $w(t)$ of the type shown in Fig. 1. At t_n, $u' > w' = t^k w = t^k u$; while at t_{n+1}, $u' < t^k u$. Hence there

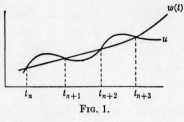

FIG. 1.

is a point s_n in between, $t_n < s_n < t_{n+1}$, at which $u' = t^k u$.

Define a principle term $T = t^a u^b (u')^c$ of $P(t,u,u')$ by means of the inequalities

(3) (a) $b + c \ge b_1 + c_1$
       (b) If $b + c = b_1 + c_1$, then $c \ge c_1$
       (c) If $b + c = b_1 + c_1$ and if $c = c_1$, then $a > a_1$

Clearly, there is only one principal term.

Now consider the ratio R of any other term in $P(t,u,u')$ to the principal term at the point s_n. We have

$$(4) \qquad \begin{aligned} |R| &= t^{a_1-a} u^{b_1-b} (u')^{c_1-c} = t^{a_1-a} u^{b_1-b} (t^k u)^{c_1-c} \\ &= t^{a_1-a+k(c_1-c)} u^{b_1+c_1-(b+c)} \end{aligned}$$

since $u' = t^k u$. If $b + c > b_1 + c_1$, then $|R| \to 0$ as $t \to \infty$ through the sequence s_n, since $u > \exp (t^{k+1}/k + 1)$, which contradicts the existence of the equation $P(t,u,u') = 0$. If $b + c = b_1 + c_1$ and if $c > c_1$, then $c - c_1 \ge 1$, and from the choice of k, we again see that $\underline{\lim} |R| = 0$

as $t \to \infty$. Finally if $b = b_1$ and if $c = c_1$, then $a > a_1$, and $\underline{\lim} |R| = 0$ once more.

If the intersections are not as pictured in Fig. 1, there must be an infinity of double roots of $u(t) - w(t) = 0$, at which points $u' = w'$, and the same contradiction is reached.

This disposes of case $(2a)$, and we turn to case $(2b)$. It is again sufficient to consider $u > w(t)$. For $t \geq t_1 \geq t_0$, one of the two inequalities, $u' - (w'u/w) > 0$ or $u' - (w'u/w) < 0$, must be valid, for

$$u' = \frac{w'u}{w}$$

infinitely often leads to the same contradiction as above.

If $u' - (w'u/w) < 0$ for $t \geq t_1$, we set $\rho(t) = u' - (w'u/w)$ and obtain, upon integrating,

$$(5) \qquad u(t) = w(t) \left[c_1 + \int_{t_1}^{t} \frac{\rho(t_1) \, dt_1}{w(t_1)} \right]$$

By assumption $\rho(t)$ is negative for $t \geq t_1$. Since $u > w > 0$, the integral $\int_{t_1}^{t} [\rho(t_1)/w(t_1)] \, dt_1$ must converge; otherwise, from (5), u would be eventually negative. Consequently there are infinitely many t for which $|\rho(t)/w(t)| \leq \epsilon$, for any $\epsilon > 0$. At these points, we have

$$(6) \qquad |u' - t^k u| \leq \epsilon w \leq \epsilon u$$

Consider the expression for $|R|$ given in (4). If $c_1 \geq c$, we use the fact that $u' < uw'/w = t^k u$ and obtain

$$(7) \qquad |R| \leq t^{a_1 - a + k(c_1 - c)} u^{b_1 + c_1 - (b + c)}$$

As before, it follows that $\lim_{t \to \infty} |R| = 0$. If $c_1 \leq c$, we use the inequality in (6), $u' - t^k u \geq -\epsilon u$ or $u' \geq (t^k - \epsilon)u$ for infinitely many points. At these points,

$$(8) \qquad |R| \leq t^{a_1 - a} (t^k - \epsilon)^{c_1 - c} u^{b_1 + c_1 - (b + c)}$$

and the proof proceeds along the same lines.

This leaves the subcase $u > w$, $u' > t^k u$, for $t \geq t_1$. Once again, consider the ratio

$$(9) \qquad |R| = t^{a_1 - a} u^{b_1 - b} (u')^{c_1 - c}$$

If $c \geq c_1$, $|R| \leq t^{a_1 - a} u^{b_1 - b} (t^k u)^{c_1 - c}$. Considering the various cases $b + c > b_1 + c_1$, $b + c = b_1 + c_1$, $c > c_1$, $b = b_1$, $c = c_1$, $a > a_1$, we see that $|R| \to 0$ as $t \to \infty$.

To treat the case where $c < c_1$, we require the following:

Lemma 1. *If $u \to \infty$, and if $u' \geq 0$ as $t \to \infty$, then $u' \leq u^{1+\epsilon}$ for $t \geq t_0$ for any $\epsilon > 0$, except perhaps in a set of intervals of finite total length which depends upon ϵ.*

Proof. Let (t_n, τ_n) be the nth interval in which $u' \geq u^{1+\epsilon}$. Then

$$(10) \qquad \int_{t_n}^{\tau_n} \frac{u' \, dt}{u^{1+\epsilon}} = \frac{1}{\epsilon} \left[\frac{1}{u(t_n)^\epsilon} - \frac{1}{u(\tau_n)^\epsilon} \right] \geq \tau_n - t_n$$

whence it follows that $\displaystyle\sum_{n=1}^{\infty} (\tau_n - t_n) < \infty$.

Now choose t to be a point outside these intervals. Then

$$(11) \qquad |R| \leq t^{a_1-a} u^{b_1-b} u^{(c_1-c)(1+\epsilon)} = t^{a_1-a} u^{b_1+c_1-(b+c)+\epsilon(c_1-c)}$$

If $b_1 + c_1 < b + c$, then $|R| \to 0$ as $t \to \infty$, provided that ϵ is chosen small enough. If $b_1 + c_1 = b + c$, we must have $c > c_1$, which is the previous case.

This concludes the proof of Theorem 1. After a thorough discussion of $u' = P(u,t)/Q(u,t)$, we shall return to $P(t,u,u') = 0$ and obtain a much more precise result, using the methods developed treating the more special equation.

3. Counterexample. One might suspect that the methods used in the proof of Theorem 1 could be utilized to obtain corresponding bounds for the solutions of equations of the form $P(t,u,u',u'') = 0$, and so on. We shall show by a simple counterexample that no such general bound can exist for second-order polynomial equations.

Theorem 2. *Let $\phi(t)$ be an arbitrary monotone increasing function of t. There exists an irrational number α such that the function*

$$(1) \qquad u(t) = \frac{1}{2 - \cos t - \cos \alpha t}$$

which is real and continuous for all t and satisfies an equation of the form $P(t,u,u',u'') = 0$, satisfies the inequality $\displaystyle\varlimsup_{t \to \infty} u(t)/\phi(t) \geq 1$.

Proof. Choose α as follows: Let (d_n) be a sequence of positive integers greater than one for which

$$(2) \quad d_r > 4\pi\phi(2\pi q_{r-1}), \qquad r = 2, 3, \ldots, q_r = d_1 d_2 \ldots d_r, r \geq 1, q_0 = 1$$

and set

$$(3) \qquad \alpha = \sum_{r=1}^{\infty} \frac{1}{q_r}$$

We see that

$$(4) \qquad \sum_{r=1}^{n} \frac{1}{q_r} = \frac{p_n}{q_n}, \qquad p_n \text{ an integer}$$

Also

$$(5) \qquad q_{n+1} > 4\pi q_n \phi(2\pi q_n)$$
$$q_{n+2} > 4\pi q_{n+1} \phi(2\pi q_{n+1}) > 4\pi q_{n+1} > (4\pi)^2 q_n \phi(2\pi q_n)$$

whence $q_{n+r} \geq (4\pi)^r q_n \phi(2\pi q_n)$. Thus

$$(6) \qquad \alpha - \frac{p_n}{q_n} = \sum_{r=n+1}^{\infty} \frac{1}{q_r} < \frac{1}{2\pi q_n \phi(2\pi q_n)}$$

If α were rational and were equal to a/b, with a and b integers, we would have

$$(7) \qquad 0 < \frac{a}{b} - \frac{p_n}{q_n} = \frac{aq_n - bp_n}{bq_n} \leq \frac{1}{2\pi q_n \phi(2\pi q_n)}$$

or

$$(8) \qquad 1 \leq aq_n - bp_n \leq \frac{b}{2\pi \phi(2\pi q_n)}$$

This is a contradiction, since $\phi(2\pi q_n) \to \infty$ as $n \to \infty$. Hence α is irrational. This implies that $2 - \cos t - \cos \alpha t$ is never zero and that $u(t)$ is continuous for all t.

For n sufficiently large

$$(9) \qquad \cos 2\pi\alpha q_n = \cos\left[2\pi q_n\left(\frac{p_n}{q_n} + \sum_{n+1}^{\infty} \frac{1}{q_r}\right)\right] = \cos\left(2\pi q_n \sum_{n+1}^{\infty} \frac{1}{q_r}\right)$$

Using (6), we see that

$$(10) \qquad 1 - \cos 2\pi\alpha q_n < \frac{1}{\phi(2\pi q_n)}$$

Hence

$$(11) \quad u(2\pi q_n) = \frac{1}{2 - \cos 2\pi q_n - \cos 2\pi\alpha q_n} = \frac{1}{1 - \cos 2\pi\alpha q_n} > \phi(2\pi q_n)$$

To show that u satisfies a second-order polynomial equation, we consider the three equations

$$(12) \qquad \begin{aligned} v &= 2 - \cos t \quad - \cos \alpha t \\ v' &= \qquad \sin t \ + \alpha \sin \alpha t \\ v'' &= \qquad \cos t + \alpha^2 \cos \alpha t \end{aligned}$$

Eliminating cos t, cos αt, sin t, and sin αt from these equations and from the two identities $\cos^2 t + \sin^2 t = 1$, $\cos^2 \alpha t + \sin^2 \alpha t = 1$, we obtain a polynomial equation in v, v', and v''. The substitution $v = 1/u$ yields the required equation in u.

4. The Solutions of $u' = P(u,t)/(Q(u,t)$. In this section, we discuss some properties of the solutions of the first-order equation

(1)
$$\frac{du}{dt} = \frac{P(u,t)}{Q(u,t)}$$

where P and Q are polynomials.

Lemma 2. *Every solution of* (1) *which is continuous for* $t \geq t_0$ *is ultimately strictly monotonic.*

Proof. It is necessary to prove that u' cannot vanish for a series of t values whose limit is ∞, unless $u \equiv c$ is a solution. The proof will be by contradiction. Let u' vanish at the points of the sequence $\{t_i\}$, $t_i \to \infty$. Then $u(t)$ and the curve defined by $P(u,t) = 0$ intersect at these points. Since P is a polynomial in u and t, the algebraic curve defined by $P = 0$ possesses only a finite number of branches, and consequently u intersects one of these branches infinitely often.

$P = 0$ and $Q = 0$ possess only a finite number of common roots, assuming, as we clearly may, that P and Q have no common factor. Consequently, for $t \geq t_1$, we may suppose that $Q(u,t)$ possesses constant sign in the immediate neighborhood of a root u_0 of $P(u,t) = 0$.

The branches of $P(u,t) = 0$ which extend to infinity as $t \to \infty$ consist of curves of the form

(2) (a) $u = c$, or
(b) $u = \phi(t)$

along which u is ultimately monotonic.

Let us examine curves of type (2b) first and show that a solution u of (1) cannot intersect any of these infinitely often. For t sufficiently large, the points of intersection cannot be maxima or minima and therefore must be points of inflection. [This is most easily seen geometrically by drawing a graph of the monotone curve $v = \phi(t)$ and the solution curve u.] But it is easy to see that, if a solution u of (1) and a branch of $P(u,t) = 0$ intersect at two successive points of inflection for u, they must also intersect at another point between these two. At this point, the slope of the solution is not zero, which contradicts the differential equation. We do not insist on filling in the rigorous details here, since we shall prove more below using a more powerful and systematic technique.

Turning to curves of type (2a), let us consider the possible intersection of $u = u(t)$, a solution of (1), and $u = c$. These intersections again must be points of inflection of $u(t)$ and have one of the four forms shown in Fig. 2. We can eliminate the forms of Fig. 2a and b immediately, since $u'(t)$ changes sign as it passes through a point of inflection, but $P(u,t)/Q(u,t)$ does not change sign in the neighborhood of $u = c$ if $u \leq c$ or $u \geq c$. If the forms of Fig. 2c and d occur, they occur only a finite number of times, for u can return to intersect $u = c$ only by intersecting one of the finite number of branches of $P(u,t) = 0$, a curve of the type $u = \phi(t)$, or a straight line $u = c_1$ in an intersection of type (2a) or (2b). This completes the proof of the lemma.

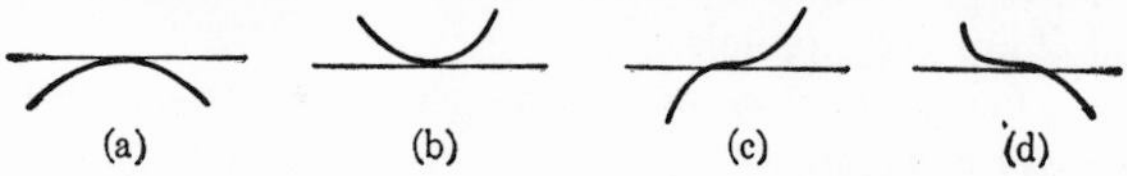

FIG. 2.

We now prove a stronger result, using a less elementary method.

Lemma 3. *If u is a solution of (1), continuous for $t \geq t_0$, then any rational function of u and t, $H(u,t) = K(u,t)/L(u,t)$ is ultimately strictly monotonic, unless $L = 0$ contains a solution of (1) or unless H is constant along a solution of (1).*

Proof. We have

$$(3) \qquad \frac{dH}{dt} = \frac{\partial H}{\partial t} + \frac{\partial H}{\partial u}\frac{P(u,t)}{Q(u,t)} = \frac{T(u,t)}{S(u,t)}$$

If dH/dt is not of constant sign as $t \to \infty$, it either vanishes infinitely often or becomes infinite infinitely often as $t \to \infty$, u traversing the curve $u = u(t)$, a solution to (1). Take first the case where it vanishes infinitely often and where consequently one branch of $T(u,t) = 0$ has an infinite number of intersections with $u(t)$. For $t \geq t_1$, this branch has an expansion of the form

$$(4) \qquad u = a_0 t^{c_0} + a_1 t^{c_1} + \cdots, \qquad c_0 > c_1 > \cdots, a_0 \neq 0$$

At the intersection with the solution

$$(5) \qquad \frac{du}{dt} = R = \frac{P(u,t)}{Q(u,t)} = b_0 t^{d_0} + b_1 t^{d_1} + \cdots, \qquad d_0 > d_1 > \cdots$$

replacing u by the series in (4). Along the branch of $T(u,t)$ given by (4),

$$(6) \qquad \frac{du}{dt} = S = a_0 c_0 t^{c_0-1} + a_1 c_1 t^{c_1-1} + \cdots$$

From the forms of R and S, we see that there are now three possibilities for large t: $R > S$, $R < S$, or $R = S$. The two inequalities hold if R and S are the slopes at the points at successive points of intersection, as one easily sees geometrically. If, on the other hand, $R = S$ for infinitely many t as $t \to \infty$, we must have $b_0 = a_0 c_0$, $b_1 = a_1 c_1$, . . . , $d_0 = c_0 - 1$, $d_1 = c_1 - 1$, and so on, whence finally $R \equiv S$. This means that $T(u,t) \equiv 0$ along a solution $u(t)$ of (1). Consequently, $dH/dt = 0$ along u, and H is constant along $u(t)$.

The case where dH/dt becomes infinite infinitely often requires that $L(u,t) = 0$ infinitely often along $u = u(t)$, since

$$(7) \qquad \frac{dH}{dt} = \frac{L \, dK/dt - K \, dL/dt}{L^2}$$

This, as above, requires that $L \equiv 0$ along $u = u(t)$.

From the above it follows that any rational function of t, u, u', . . . is ultimately strictly monotonic, apart from the trivial exceptions mentioned above.

5. Asymptotic Behavior of Solutions of $u' = P(u,t)/Q(u,t)$. We are now able to prove the following remarkable result:

Theorem 3. *Any solution of the equation*

$$(1) \qquad \frac{du}{dt} = \frac{P(u,t)}{Q(u,t)}$$

continuous for $t \geq t_0$, is ultimately monotonic, together with all its derivatives, and satisfies one or the other of the relations

$$(2) \qquad u \sim at^b e^{P(t)}, \qquad u \sim at^b (\log t)^{1/c}$$

where $P(t)$ is a polynomial in t and c is an integer.

Proof. Consider $Q(u,t)u' - P(u,t) = 0$, the terms of which are of the form $a_1 t^m u^n$, or $b_1 t^m u^n u'$. Since all rational functions in u, u', and t are ultimately monotonic, the ratio of any two such terms approaches a limit as $t \to \infty$. This limit may be $\pm \infty$, 0, or a nonzero constant, but there must be one quotient of two terms which approaches a nonzero constant.

If one of the two terms contains u', but not the other, we obtain

$$(3) \qquad u' u^n t^m \sim c_1$$

If both or neither contain u', the result is

$$(4) \qquad u \sim c_2 t^{p/q}$$

where p/q is a rational fraction.

The first case presents different results according to the subcases:

$$
\begin{aligned}
(5) \quad &(a) \ n \neq -1, \ m \neq +1 \\
&(b) \ n = -1, \ m \neq +1 \\
&(c) \ n = -1, \ m = +1 \\
&(d) \ n \neq -1, \ m = +1
\end{aligned}
$$

Corresponding to the various cases in (5) we have the respective asymptotic behaviors:

$$(6) \quad (a) \quad \frac{u^{n+1}}{n+1} \sim c_1 \frac{t^{1-m}}{1-m} + d_1$$

$$(b) \quad \log u \sim c_1 \frac{t^{1-m}}{1-m} + d_1$$

$$(c) \quad \log u \sim c_1 \log t$$

$$(d) \quad \frac{u^{n+1}}{n+1} \sim c_1 \log t$$

Cases $(6a)$ and $(6d)$ are in the form stated by the theorem; cases $(6b)$ and $(6c)$ are not, and a further discussion is required. We use the method we have previously applied to derive more precise results, namely, substitute the crude result in the differential equation, and use the equation again. Let us consider case $(6b)$ first.

We write the equation in the form

$$(7) \qquad \frac{du}{dt} = \frac{P_0 u^r + P_1 u^{r-1} + \cdots + P_r}{Q_0 u^s + Q_1 u^{s-1} + \cdots + Q_s}$$

where P_k and Q_l are polynomials in t. We may take $1 - m$ as positive, since if it is negative, $u \sim 1$. If $1 - m > 0$ and if $c_1 > 0$, we see that in (7) $r = s + 1$. Hence dividing through by u^s, we obtain

$$(8) \qquad \frac{du}{dt} = \frac{P_0}{Q_0} u + R_1(t) + O\left(\frac{t^a}{u}\right)$$

for some a, or

$$(9) \qquad \frac{du/dt}{u} = \frac{P_0}{Q_0} + O\left(\frac{1}{t^2}\right)$$

for large t. Integrating,

$$(10) \qquad \log u = P(t) + c_3 \log t + O\left(\frac{1}{t}\right)$$

The case $c_1 < 0$ may be treated by replacing u by $1/u$ in the original equation.

We turn now to the last case, case $(6c)$, which requires the consideration of a great many subcases. There are two terms, $at^b u^c u'$ and $dt^{b-1} u^{c+1}$, of equal order. Further we may assume that there is no other term of equal order, since the contrary assumption yields the relation $u \sim c_4 t^{c_5}$. If T is any third term, the quotient

$$(11) \qquad \frac{(at^b u^c u' - dt^{b-1} u^{c+1})}{T}$$

tends to a limit as $t \to \infty$. There are now two possibilities:

(*a*) There is a third term whose order is equal to that of the difference between the two principal terms

(*b*) There is no such third term

Let us consider the first possibility. In this case we have a relation either of the form

$$(12) \qquad at^b u^c u' - dt^{b-1} u^{c+1} \sim et^f u^g$$

or of the form

$$(13) \qquad at^b u^c u' - dt^{b-1} u^{c+1} \sim et^f u^g u'$$

Take the case (12) first. From (6c) we obtain the crude result $u = t^{c_1+\epsilon}$, where $\epsilon = \epsilon(t) \to 0$ as $t \to \infty$, and where $c_1 = d/a$. Set $u = t^{d/a} v$. Substituting, (12) becomes

$$(14) \qquad v^{c-g} v' \sim \frac{e}{a} t^{f + \frac{dg}{a} - \frac{d(c+1)}{a-b}}$$

Recalling that $\log u \sim (d \log t)/a$ implies that $\log v = O(\log t)$ as $t \to \infty$, we see by an enumeration of cases that U has one or the other of the forms in (2).

The case where (13) holds and the case where possibility (*b*) holds we leave as exercises for the reader.

6. A Sharpening of Theorem 1. We are now able to prove a much sharper form of Theorem 1, namely,

Theorem 4. *Let u be any solution of the polynomial equation*

$$P(t,u,u') = 0$$

continuous for $t \geq t_0$. Then either

$$(1) \qquad u = o(t^b)$$

for some b, or

$$(2) \qquad u = \exp \left[at^b (1 + \epsilon(t)) \right]$$

where a and b are fixed constants and $\epsilon(t) \to 0$ as $t \to \infty$.

All solutions of the latter class are monotonic, together with all their derivatives.

Proof. Let us assume that no constant b exists for which $u = o(t^b)$. Then, however large we choose b, it is possible to find values of t such that $u \geq t^b$. Choose an increasing sequence $b_v \to \infty$, and a sequence t_v such that

$$(3) \qquad u(t_v) \geq t_v^{b_v}$$

Let us now construct a curve $u = t^{b(t)} = e^{b(t)\log t} = e^{\phi(t)}$ passing through the points $(t_v, t_v^{b_v})$, and possessing the following properties:

(a) $b'(t)$ is positive and continuous
(b) $b(t)/t^c \to 0$ as $t \to \infty$ for any $c > 0$
(c) $b'(t)/t^{1-c} \to 0$ for any $c > 0$

This can always be done by taking $b_v = v$ and choosing the points t_v sufficiently far apart. In terms of $\phi(t)$, we have, for any $c > 0$,

(4) $$t\phi' \to \infty, \qquad t^{1-c}\phi' \to 0$$

We shall now show that the statement $u(t) > e^{\phi(t)}$ for a solution of $P(t,u,u') = 0$ for infinitely many $t \geq t_0$ implies that the inequality holds

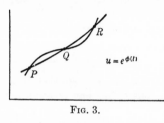

FIG. 3.

for all t. The proof is by contradiction. If the inequality does not hold for all $t \geq t_0$, there are infinitely many intersections of the two curves $n = u(t)$ and $u = e^{\phi(t)}$, as in Fig. 3. The argument is considerably simplified if there are an infinity of points of contact, so there is no loss of generality in considering the above situation.

As in previous discussions in Sec. 2, we see that there is a point between P and Q for which

(5) $$u' = \phi'u, \qquad u \geq e^{\phi}$$

u here representing the solution curve. Using the concept of "principal term" in $P(t,u,u')$ developed above, we readily obtain a contradiction, combining (4) and (5). Consequently, the inequality $u > e^{\phi(t)}$ must hold for all sufficiently large t.

In this case any expression of the type $H = t^a u^b (u')^c$, with $c \neq 0$, is ultimately monotonic. To show this, eliminate u' between $t^a u^b (u')^c = H$ and the relation $P(t,u,u') = 0$.

The result is $F(t,u,H) = 0$. From this we obtain, upon differentiation,

(6) $$\frac{\partial F}{\partial t} + \frac{\partial F}{\partial u}u' + \frac{\partial F}{\partial H}\frac{dH}{dt} = 0$$

If $dH/dt = 0$, we must have $\partial F/\partial t + (\partial F/\partial u)u' = 0$. Substituting this value for u' in $P(t,u,u') = 0$, we obtain a polynomial relation between u, t, and H. Eliminating H between this relation and $F(t,u,H) = 0$, the resulting relation is a polynomial equation between u and t. This, however, is a contradiction of the property $u > e^{\phi(t)}$ for $t \geq t_0$.

Now choose two terms of equal order, as $t \to \infty$, in $P(t,u,u')$:

$$T_1 = t^{a_1} u^{b_1} (u')^{c_1}$$

and $T_2 = t^{a_2} u^{b_2} (u')^{c_2}$. Since $u > e^{\phi(t)}$, c_1 cannot equal c_2; for if c_1 should equal c_2, the relation $T_1/T_2 \sim d$ as $t \to \infty$ would imply $u \sim c_3 t^{c_4}$, which contradicts $u > e^{\phi(t)}$. The relation $T_1/T_2 \sim d$ then takes the form

$$(7) \qquad\qquad u' u^\alpha \sim c_4 t^\beta, \qquad c_4 \neq 0$$

and the relation $u > e^{\phi(t)}$ forces α to be -1. From this, integration yields

$$(8) \qquad\qquad u = \exp\left[c_s t^\beta (1 + \epsilon(t))\right]$$

where $\epsilon(t) \to 0$ for $t \to \infty$. More precise asymptotic expressions for u may now be obtained using the type of argument presented in Sec. 5.

7. Some Results Concerning $u'' = R(u,t)/Q(u,t)$. As we know from the counterexample given in Sec. 3, any general theory of the asymptotic behavior of proper solutions of polynomial equations of the form

$$P(t, u, u', u'') = 0$$

will be extremely difficult. If, however, we restrict our attention to the important class of equations having the form

$$(1) \qquad\qquad u'' = \frac{P(u,t)}{Q(u,t)}$$

where P and Q are polynomials, we can obtain some interesting results. The proofs of these results employ the same techniques as above, and for this reason we shall leave them as exercises.

Exercises

1. If u is a proper solution of (1), then either there exists a constant k such that $u = O(t^k)$, or there exist two constants a and b such that $u = \exp a t^{b+\epsilon}$, where $\epsilon = \epsilon(t) \to 0$ as $t \to \infty$.

2. If $u'' = P(u,t)$, where P is a polynomial of degree greater than one, then $u = O(t^k)$ for some k if u is a proper solution.

3. If u is a proper solution of (1), then $u = O(t^k)$ if the degree of P in u does not exceed the degree of Q in u by unity.

4. If u is a solution of (1) with the property that any rational function of u', u, and t is ultimately monotonic, then as $t \to \infty$, u has one of the following forms:

(a) $\exp\left[(a + \epsilon)t^b e^{p(t)}\right]$ $\qquad$ (b) $\exp\left[(a + \epsilon)(t^q \log t)^{1/p}\right]$

(c) $\exp\left[(a + \epsilon)t^b\right]$ $\qquad\qquad$ (d) $\exp\left[(a + \epsilon)(\log t)^{(p+1)/p}\right]$

(e) $a(\log \log t)^{1/p}$ $\qquad\qquad$ (f) $a(\log t)^p (\log \log t)^{1/q}$

(g) $a(\log t)^b$ $\qquad\qquad\qquad$ (h) $\exp\left[(a + \epsilon)(\log t)^{(p-1)/p}\right]$

where, whenever ϵ appears, it is understood to be a function of t, which approaches zero as $t \to \infty$, and where p and q are integers, a and b are real numbers, and $p(t)$ in (a) is a polynomial.

BIBLIOGRAPHY

Section 1.   For the application of the theory of functions of a complex variable to the structure of the solutions of first-order differential equations, see:

Boutroux, P., *Leçons sur les fonctions définies par les équations différentielles du premier ordre*, Gauthier-Villars & Cie, Paris, 1908.

Section 2

Borel, E., *Mémoire sur les séries divergents*, Ann. École Norm., vol. 16 (1899), pp. 9–136, pp. 26*ff*.

Lindelof, E., *Sur la croissance des intégrales des équations différentielles algébriques du premier ordre*, Bull. Soc. Math. France, vol. 27 (1899), pp. 205–215.

Section 3

Vijayaraghavan, T., N. M. Basu, and S. N. Bose, *A simple example for a theorem of Vijayaraghavan*, J. London Math. Soc., vol. 12 (1937), pp. 250–252.

Sections 4, 5, and 6

Hardy, G. H., *Some results concerning the behavior at infinity of a real and continuous solution of an algebraic differential equations of the first order*, Proc. London Math. Soc., ser. 2, vol. 10 (1912), pp. 451–468.

Section 7

Fowler, R. H., *Some results on the form near infinity of real continuous solutions of a certain type of second order differential equation*, Proc. London Math. Soc., ser. 2, vol. 13 (1914), pp. 341–371.

CHAPTER 6

THE SECOND-ORDER LINEAR DIFFERENTIAL EQUATION

1. Introduction. We now turn our attention to the second-order linear differential equation

$$(1) \qquad \frac{d}{dt}\left(k(t)\,\frac{du}{dt} \right) + l(t)u = 0$$

Since it is virtually impossible to present a complete account of all that is known concerning the properties of the solutions of the above equation, we have rather attempted to present a cross section of theorems and techniques in such a fashion that the reader will have little difficulty in following original papers or in deriving new results.

Although some of the results we state below are special cases of general theorems valid for linear equations of any order, the results for the most part depend very strongly upon the particularly simple form of (1). Even where this duplication exists, we shall not hesitate to present a proof applicable only to (1) if it illustrates an important and useful technique.

The physical importance of equations of the above class can hardly be overestimated, and this accounts for the vast amount of research connected with (1). Mathematically, the equation presents a continual challenge to the skill of the analyst to extract as many properties of the solution as possible without the luxury of an explicit representation for u in terms of the coefficients k and l.

We shall begin our discussion with some preliminary lemmas required in what follows. Then we shall turn to the questions of boundedness, oscillation, and asymptotic behavior of the solutions.

2. Some Lemmas. In this section we gather together some results we shall call upon repeatedly below. Lemma 1 has already been stated and proved in Chap. 2, and Lemma 2 is a particularization of Theorem 3 of Chap. 1.

Lemma 1. *Let* $u,v \geq 0$, $c_1 > 0$, *and* u *satisfy the inequality*

$$(1) \qquad u \leq c_1 + \int_0^t uv\,\dot{dt}, \qquad t \geq 0$$

Then

$$u \le c_1 \exp \left(\int_0^t v \, dt \right) \tag{2}$$

Lemma 2. *Let u_1 and u_2 be two linearly independent solutions of*

$$u'' - a(t)u = 0 \tag{3}$$

for which the Wronskian

$$w = \begin{vmatrix} u_1 & u_2 \\ u_1' & u_2' \end{vmatrix} = 1 \tag{4}$$

for all t. Then the general solution of the inhomogeneous equation

$$u'' - a(t)u = w(t) \tag{5}$$

is given by

$$u = c_1 u_1 + c_2 u_2 + \int_0^t [u_1(t)u_2(t_1) - u_1(t_1)u_2(t)]w(t_1) \, dt_1 \tag{6}$$

where c_1 and c_2 are constants determined by the initial conditions.

Lemma 3. *If u_1 is a solution of* (3), *then*

$$u_2 = u_1 \int^t \frac{dt}{u_1^2} \tag{7}$$

is another, linearly independent, solution.

Proof. The result may be obtained by the standard variation-of-parameters method, letting $u_2 = u_1 v$, or by observing that for any two solutions of (3), u_1 and u_2, we have (compare Theorem 2, Chap. 1)

$$w = \begin{vmatrix} u_1 & u_2 \\ u_1' & u_2' \end{vmatrix} = c_1 \tag{8}$$

This equation is a first-order equation for u_2, which may be solved easily to yield (7).

3. Some Useful Transformations. In this section, we consider some changes of independent and dependent variable which will be of great service in obtaining properties of the solutions and in reducing the number of distinct types of equations.

Let us begin by demonstrating two methods by means of which

$$\frac{d}{dt} \left(k(t) \frac{du}{dt} \right) + l(t)u = 0 \tag{1}$$

may be reduced to the simpler form

$$\frac{d^2u}{dt^2} + a(t)u = 0 \tag{2}$$

The first method depends upon a change of independent variable and assumes that $k(t)$ is eventually positive and that

$$(3) \qquad \int^{\infty} \frac{dt}{k(t)} = \infty$$

For this case, set $s = \int^{t} dt/k(t)$.   Equation (1) becomes

$$(4) \qquad \frac{d^2u}{ds^2} + k(t)l(t)u = 0$$

where $k(t)l(t)$ is now a function of s.   As $t \to \infty$, $s \to \infty$.

The second method depends upon a change of dependent variable. Write (1) in its full expansion as

$$(5) \qquad u'' + \frac{k'(t)}{k(t)}\, u' + \frac{l(t)}{k(t)}\, u = 0$$

We now wish to transform (5) into an equation lacking a middle term. That this may always be accomplished is the substance of the following:

Lemma 4.   *The substitution*

$$(6) \qquad u = v \exp\left(-\tfrac{1}{2} \int_0^t p\, dt\right)$$

transforms

$$(7) \qquad u'' + p(t)u' + q(t)u = 0$$

into

$$(8) \qquad v'' + \left(q - \frac{1}{2}\, p' - \frac{p^2}{4}\right)v = 0$$

An important feature of this transformation is that, provided $\int_0^t p\, dt$ is finite for finite t, the zeros of v are the same as the zeros of u.

Applying Lemma 4 to (5), the substitution $v = u\sqrt{k(t)}$ yields

$$(9) \qquad v'' + \left[\frac{l(t)}{k(t)} - \frac{1}{2}\frac{d}{dt}\left(\frac{k'}{k}\right) - \frac{(k'/k)^2}{4}\right]v = 0$$

an equation of type (2).

We now wish to reduce (2) to a form which in many cases is easier to treat.

Lemma 5 (The Liouville Transformation).   *The change of variable*

$$(10) \qquad s = \int_0^t a(t)\, dt$$

transforms

$$(11) \qquad u'' \pm a^2(t)u = 0$$

into

$$(12) \qquad \frac{d^2u}{ds^2} + \frac{a'(t)}{a^2(t)} \frac{du}{ds} \pm u = 0$$

Using Lemma 4 and letting

$$(13) \qquad v = u\sqrt{a(t)} \left[= u \exp\left(\frac{1}{2}\int \frac{a'(t)\,ds}{a^2(t)}\right) \right]$$

the resultant equation for v is

$$(14) \qquad \frac{d^2v}{ds^2} + \left[\pm 1 - \frac{1}{2}\frac{d}{ds}\left(\frac{a'(t)}{a^2(t)}\right) - \frac{1}{4}\left(\frac{a'(t)}{a^2(t)}\right)^2 \right] v = 0$$

In many cases, first-order equations are easier to discuss than second-order equations. In general, we can always reduce an nth-order linear differential equation to an $(n-1)$st-order nonlinear differential equation by means of the transformation $u'/u = v$. In the case $n = 2$, the result is particularly simple.

Lemma 6. *The substitution*

$$(15) \qquad u = \exp \int^t v\, dt$$

transforms (2) *into*

$$(16) \qquad v' + v^2 + a(t) = 0$$

Equations of type (16) are called Riccati equations. It is easy to see that (15) establishes a connection between the general second-order equation

$$(17) \qquad u'' + p(t)u' + q(t)u = 0$$

and the general Riccati equation

$$(18) \qquad u' = a(t)u^2 + b(t)u + c(t)$$

Exercise

Find the second-order linear equation equivalent to (18).

────────────

It should be pointed out that, in all cases where these changes of variable are employed, care must be exercised in verifying the one-to-one nature of the transformation.

Finally we note that any second-order equation

$$(19) \qquad u'' + p(t)u' + q(t)u = 0$$

is equivalent to a second-order system, namely,

$$(20) \qquad \begin{aligned} u' &= v \\ v' &= -p(t)v - q(t)u \end{aligned}$$

It is sometimes convenient to introduce polar coordinates in place of cartesian. Taking the general second-order system

$$(21) \qquad \begin{aligned} u' &= a_{11}(t)u + a_{12}(t)v \\ v' &= a_{21}(t)u + a_{22}(t)v \end{aligned}$$

we may state

Lemma 7. *The change of variable*

$$(22) \qquad \begin{aligned} u &= \rho \cos \theta \\ v &= \rho \sin \theta, \qquad \rho > 0 \end{aligned}$$

transforms (21) *into*

$$(23) \qquad \theta' = \frac{a_{21} - a_{12}}{2} + \frac{1}{2} r(t) \cos (2\theta + \psi)$$

$$\frac{\rho'}{\rho} = \frac{a_{11} + a_{22}}{2} + \frac{1}{2} r(t) \sin (2\theta + \psi)$$

where

$$(24) \qquad r = \sqrt{(a_{11} - a_{22})^2 + (a_{12} + a_{21})^2}$$

$$\cos \psi = \frac{a_{21} + a_{12}}{r}$$

$$\sin \psi = \frac{a_{11} - a_{22}}{r}$$

4. Boundedness Theorems. Having disposed of these preliminaries, let us now turn to the problem of determining when all solutions of a given equation are bounded as $t \to \infty$. We begin by considering the equation

$$(1) \qquad u'' + (a^2 + \phi(t))u = 0$$

where $\phi(t) \to 0$ as $t \to \infty$. Without loss of generality, we may assume that $a = 1$.

The question naturally arises as to the connection between the solutions of (1) and those of the easily soluble equation

$$(2) \qquad u'' + u = 0$$

We shall show that in most "ordinary" cases, there is a close correspondence, but that the condition $\phi(t) \to 0$ as $t \to \infty$ is by no means sufficient to ensure the boundedness of all solutions of (1).

Our first result is

Theorem 1. *All solutions of*

$$(3) \qquad u'' + (1 + \phi(t) + \psi(t))u = 0$$

are bounded, provided that

(4) (a) $\displaystyle\int^{\infty} |\phi(t)| \, dt < \infty$

    (b) $\displaystyle\int^{\infty} |\psi'(t)| \, dt < \infty$, $\psi(t) \to 0$ *as* $t \to \infty$

Proof. We show first that the equation

$$(5) \qquad u'' + (1 + \psi(t))u = 0$$

has all solutions bounded provided that (4b) is fulfilled. From (5), multiplying by u' and integrating between 0 and t, we obtain

$$(6) \qquad \frac{u'^2}{2} + \frac{u^2}{2} + \int_0^t \psi(t_1)uu' \, dt_1 = c_1$$

Integrating by parts, this yields

$$(7) \qquad \frac{u'^2}{2} + \frac{u^2}{2} + \psi(t)\frac{u^2}{2} - \frac{1}{2}\int_0^t \psi'(t_1)u^2 \, dt_1 = c_2$$

Take t large enough so that $1 + \psi(t) \geq \frac{1}{2}$. Then for $t \geq t_0$,

$$(8) \qquad u^2 \leq 4|c_2| + 2\int_0^t |\psi'(t_1)|u^2 \, dt_1 = c_3 + 2\int_0^t |\psi'(t_1)|u^2 \, dt_1$$

Applying Lemma 1, the result is, for $t \geq t_0$,

$$(9) \qquad u^2 \leq c_3 \exp\left(2\int_{t_0}^t |\psi'(t_1)| \, dt_1\right) \leq c_3 \exp\left(2\int_{t_0}^{\infty} |\psi'(t_1)| \, dt_1\right)$$

Hence u is bounded. To finish the proof, we must demonstrate the following:

Theorem 2. *If all solutions of*

$$(10) \qquad u'' + a(t)u = 0$$

are bounded, then all solutions of

$$(11) \qquad u'' + (a(t) + b(t))u = 0$$

are also bounded, if

$$(12) \qquad \int^{\infty} |b(t)| \, dt < \infty$$

This, however, is a particular case of Theorem 6 of Chap. 2, as we see upon transforming (11) into a system.

Exercise

Use Lemma 7 to show that all solutions of $u'' + (1 + f(t))u = 0$ are bounded if $\int^{\infty} |f(t)| \, dt < \infty$.

---

5. A Counterexample. We now show by an example how close Theorem 1 is to being best possible in one sense.

Theorem 3. *If, as $t \to \infty$,*

$$(1) \qquad\qquad g(t) \to 0, \qquad g'(t) \to 0$$

then

$$(2) \qquad\qquad u = \left[\exp \left(\int_0^t g(s) \cos s \, ds \right) \right] \cos t$$

is a solution of the equation

$$(3) \qquad\qquad u'' + (1 + \phi(t))u = 0$$

where

$$(4) \qquad \phi(t) = 3g(t) \sin t - g'(t) \cos t - g^2(t) \cos^2 t$$

approaches zero as $t \to \infty$.

Choosing $g(t) = (\cos t)/t$, we see that we obtain unbounded solutions, although $\phi(t)$ and $\phi'(t)$ are both $O(1/t)$ as $t \to \infty$.

6. $u'' + a(t)u = 0$, $a(t) \to \infty$. In the previous sections we have examined the case where $a(t) \to a^2 \neq 0$ as $t \to \infty$. Here we consider the case where $a(t) \to \infty$.

Theorem 4. *If $a(t) \to \infty$ monotonically, all solutions of*

$$(1) \qquad\qquad u'' + a(t)u = 0$$

are bounded as $t \to \infty$.

Proof. We have

$$(2) \qquad\qquad u'u'' + a(t)uu' = 0$$

Integrating between 0 and t, and then by parts,

$$(3) \qquad\qquad \frac{u'^2}{2} + a(t)\frac{u^2}{2} - \int_0^t \frac{u^2}{2} \, da(t) = c_1$$

whence

$$(4) \qquad\qquad a(t)\frac{u^2}{2} \leq |c_1| + \int_0^t \frac{u^2}{2} \, da(t)$$

$$\leq |c_1| + \int_0^t \frac{u^2 a(t)}{2} \, \frac{da(t)}{a(t)}$$

Applying Lemma 1,

$$(5) \qquad \frac{a(t)u^2}{2} \le |c_1| \exp\left(\int_0^t \frac{da(t)}{a(t)}\right) \le |c_1|a(t)$$

Consequently $u^2 \le 2|c_1|$.

7. $u'' + a(t)u = 0$, $a(t) \to 0$. We have considered the cases where $a(t) \to a^2 \ne 0$ and where $a(t) \to \infty$ as $t \to \infty$. Let us now discuss the case $a(t) \to 0$ as $t \to \infty$. Consideration of the equation

$$(1) \qquad u'' + \frac{u}{t^2} = 0$$

which has two linearly independent solutions of the form t^{α_1} and t^{α_2}, where α_1 and α_2 are the roots of $\alpha^2 - \alpha + 1 = 0$, shows that solutions of equations of this type can be unbounded as $t \to \infty$. It is reasonable to expect that, if $a(t) \to 0$ sufficiently rapidly as $t \to \infty$, the solutions of $u'' + a(t)u = 0$ will approach those of $u'' = 0$.

Theorem 5. *Consider the equation*

$$(2) \qquad u'' + a(t)u = 0$$

where

$$(3) \qquad \int^{\infty} t|a(t)|\, dt < \infty$$

Then $\lim\limits_{t \to \infty} u'$ *exists, and the general solution is asymptotic to* $d_0 + d_1 t$ *as* $t \to \infty$, *where* d_1 *may be zero, or* d_0 *may be zero, but not both simultaneously.*

Proof. Write $u'' = -a(t)u$. Integrating twice between 1 and t, we obtain

$$(4) \qquad u = c_1 + c_2 t - \int_1^t (t - t_1)a(t_1)u(t_1)\, dt_1$$

From this we obtain, for $t \ge 1$,

$$(5) \qquad |u| \le (|c_1| + |c_2|)t + t\int_1^t |a(t_1)|\, |u(t_1)|\, dt_1$$

or

$$(6) \qquad \frac{|u|}{t} \le |c_1| + |c_2| + \int_1^t t_1|a(t_1)| \frac{|u(t_1)|}{t_1}\, dt_1$$

Applying our fundamental lemma, we derive, for $t \ge 1$,

$$(7) \qquad \frac{|u|}{t} \le (|c_1| + |c_2|) \exp\left[\int_1^t t_1|a(t_1)|\, dt_1\right] \le c_3$$

Returning to (4), differentiation yields

$$(8) \qquad u' = c_2 - \int_1^t a(t_1)u(t_1) \, dt_1$$

Since $\int_1^t a(t_1)u(t_1) \, dt_1$ is majorized by

$$\int_1^t |a(t_1)| \, |u(t_1)| \, dt_1 \le c_3 \int_1^t t_1 |a(t_1)| \, dt_1$$

the infinite integral converges, and u' has a limit as $t \to \infty$.

If this limit is not zero, we see that $u \sim d_1 t$, with $d_1 \ne 0$, as $t \to \infty$. Using the fact that $v = u \int_t^\infty dt/u^2$ is another solution, we derive a solution v which is asymptotic to 1 as $t \to \infty$.

To ensure that this limit is not zero, we choose $c_2 = 1$ and use as a lower limit, instead of 1, a point t_0, where t_0 is chosen so that

$$1 - c_3 \int_{t_0}^\infty t_1 |a(t_1)| \, dt_1 > 0$$

This theorem completes our preliminary discussion of the boundedness of the solutions of $u'' + a(t)u = 0$. In later sections we shall obtain more precise results.

Exercises

1. What is the related result for $d^n u/dt^n + a(t)u = 0$?

2. Is there an analogous theorem for the vector equation

$$dy/dt = A(t)y$$

3. Show that, if $d^n u/dt^n + \sum_{k=0}^{n-1} a_{n-k}(t) \, d^k u/dt^k = 0$ and if

$$\int_0^\infty |a_k(t)| t^{k-1} \, dt < \infty$$

then $\lim_{t \to \infty} d^{n-1} u/dt^{n-1}$ exists.

---

8. L^2 Boundedness. In the previous sections we have been concerned with the boundedness of the solutions of $u'' + a(t)u = 0$, using the norm $\|u\| = \varlimsup_{0 \le t < \infty} |u|$. In this section we consider a different norm, which is of interest in connection with various problems of mathematical physics, namely,

$$(1) \qquad \|u\| = \left(\int_0^\infty |u^2(t)| \, dt \right)^{1/2}$$

If the norm is finite, we say that u belongs to $L^2(0, \infty)$, written $u \in L^2(0, \infty)$.

Theorem 6. *If all the solutions of $u'' + a(t)u = 0$ belong to $L^2(0, \infty)$, then all the solutions of*

$$(2) \qquad u'' + (a(t) + b(t))u = 0$$

belong to $L^2(0, \infty)$, if $|b(t)| \leq c_1$, $t \geq 0$.

Proof. Let u_1 and u_2 be two linearly independent solutions of

$$u'' + a(t)u = 0$$

such that

$$(3) \qquad w = \begin{vmatrix} u_1 & u_2 \\ u_1' & u_2' \end{vmatrix} = 1$$

for all t. Utilizing Lemma 2, any solution of (2) satisfies an integral equation of the form

$$(4) \qquad u = c_2 u_1 + c_3 u_2 - \int_0^t [u_1(t)u_2(t_1) - u_1(t_1)u_2(t)]b(t_1)u(t_1) \, dt_1$$

From this we obtain

$$(5) \qquad |u| \leq |c_2| \, |u_1| + |c_3| \, |u_2| \\ + \int_0^t [|u_1(t)| \, |u_2(t_1)| + |u_1(t_1)| \, |u_2(t)|]|b(t_1)| \, |u(t_1)| \, dt_1$$

We now require the following lemma (Cauchy-Schwarz inequality):

Lemma 8. *For all f and g for which the right side exists, we have*

$$(6) \qquad \int_a^b fg \, dt \leq \left(\int_a^b |f|^2 \, dt \right)^{\frac{1}{2}} \left(\int_a^b |g|^2 \, dt \right)^{\frac{1}{2}}$$

Proof of Lemma. Since $(u - v)^2 \geq 0$, we have

$$(7) \qquad 2uv \leq u^2 + v^2$$

Set $u = f / \left(\int_a^b f^2 \, dt \right)^{\frac{1}{2}}$ and $v = g / \left(\int_a^b g^2 \, dt \right)^{\frac{1}{2}}$, integrate both sides of (7) between a and b, and simplify.

Returning to (5), we have, applying the consequence of $(u + v)^2 \leq 2(u^2 + v^2)$, that

$$(8) \quad u^2 \leq 4 \left\{ c_2^2 u_1^2 + c_3^2 u_2^2 + \left[\int_0^t (|u_1(t)| \, |u_2(t_1)| \right. \right. \\ \left. \left. + |u_1(t_1)| \, |u_2(t)|)|b(t_1)| \, |u(t_1)| \, dt_1 \right]^2 \right\}$$

Using the inequality of (6) and then $(u + v)^2 \leq 2(u^2 + v^2)$ on the last integral, we obtain

$$(9) \quad u^2 \leq 8 \left\{ c_2^2 u_1^2 + c_3^2 u_2^2 + \left[u_1^2(t) \int_0^t u_2^2(t_1) \, dt_1 \right. \right. \\ \left. \left. + u_2^2(t) \int_0^t u_1^2(t_1) \, dt_1 \right] \left[\int_0^t b^2(t_1)u^2(t_1) \, dt_1 \right] \right\}$$

which, by hypothesis, simplifies to

(10) $$u^2 \leq 8\left[c_2^2 u_1^2 + c_3^2 u_2^2 + c_4(u_1^2 + u_2^2) \int_0^t u^2(t_1)\, dt_1 \right]$$

Integrating both sides between 0 and t, we derive

(11) $$\int_0^t u^2\, dt_1 \leq 8c_2^2 \int_0^t u_1^2\, dt_1 + 8c_3^2 \int_0^t u_2^2\, dt_1$$
$$+ 8c_4 \int_0^t (u_1^2 + u_2^2)\left(\int_0^{t_2} u^2(t_1)\, dt_1 \right) dt_2$$
$$\leq c_5 + 8c_4 \int_0^t \left[(u_1^2 + u_2^2)\left(\int_0^{t_2} u^2(t_1)\, dt_1 \right) \right] dt_2$$

Applying our principal inequality, this becomes

(12) $$\int_0^t u^2\, dt_1 \leq c_5 \exp\left[8c_4 \int_0^t (u_1^2 + u_2^2)\, dt \right]$$

whence $u \in L^2(0, \infty)$.

Exercises

1. Show, by comparing areas, that

$$uv \leq \frac{u^p}{p} + \frac{v^{p'}}{p'}$$

for u and $v \geq 0$, where $p > 1$ and $p' = p/(p-1)$.

2. (Hölder's inequality.) Use the above inequality to show that

$$\smallint uv\, dt \leq (\smallint u^p\, dt)^{1/p}(\smallint v^{p'}\, dt)^{1/p'}$$

for u and $v \geq 0$, where p and p' are as above.

3. Show that, if all solutions of $u'' + a(t)u = 0$ belong to $L^p(0, \infty)$ and $L^{p'}(0, \infty)$, then the same is true of the solutions of $u'' + (a(t) + b(t))u = 0$, provided that $|b(t)| \leq c_1$ for $t \geq t_0$.

4. Can all solutions of $u'' + a(t)u = 0$ belong to $L(0, \infty)$ and be bounded?

5. Show that, if all solutions of $u'' + a(t)u = 0$ belong to $L(0, \infty)$ and are bounded, the same is true of the solution of $u'' + (a(t) + b(t))u = 0$ if $|b(t)| \leq c_1$ for $t \geq t_0$.

---

9. Relations between $\|u\|$, $\|u'\|$, and $\|u''\|$. Let us now consider what types of inequalities exist between the norms of various derivatives, considering only the two most common norms

(1) (a) $\|u\| = \varlimsup\limits_{0 \leq t < \infty} |u|$

(b) $\|u\| = \left(\int_0^\infty u^2\, dt \right)^{1/2}$

There has been a considerable amount of research on these equations, and the method we present below is perhaps not the most efficient. It is, however, interesting and applicable, equally, to many different types of norms.

Theorem 7. *Using either of the above norms, the boundedness of $\|u\|$ and $\|u''\|$ implies that of $\|u'\|$.*

Proof. Let $\|u\|$ and $\|u''\|$ be finite and let $f(t)$ be defined by the equation

$$(2) \qquad u'' - u = f(t)$$

By hypothesis, then, $\|f(t)\|$ is finite. Considering u as a solution of the second-order linear differential equation, u is given by

$$(3) \quad u = c_1 e^t + c_2 e^{-t} + \tfrac{1}{2} \int_0^t (e^{t-t_1} - e^{-(t-t_1)}) f(t_1)\, dt_1$$
$$= e^t \left(c_1 + \tfrac{1}{2} \int_0^t e^{-t_1} f(t_1)\, dt_1 \right) + e^{-t} \left(c_2 - \tfrac{1}{2} \int_0^t e^{t_1} f(t_1)\, dt_1 \right)$$

With reference to either norm, $\int_0^\infty e^{-t_1} f(t_1)\, dt_1$ is convergent if $\|f\|$ is finite. Furthermore, it is easy to verify that $e^{-t} \int_0^t e^{t_1} f(t_1)\, dt_1$ has bounded norm if $\|f\|$ is finite. Hence, if u is to have bounded norm, it is necessary that

$$(4) \qquad c_1 + \tfrac{1}{2} \int_0^\infty e^{-t_1} f(t_1)\, dt_1 = 0$$

Using this relation, we may write

$$(5) \qquad u = -\frac{e^t}{2} \int_t^\infty e^{-t_1} f(t_1)\, dt_1 + c_2 e^{-t} - \frac{e^{-t}}{2} \int_0^t e^{t_1} f(t_1)\, dt_1$$

whence

$$(6) \qquad u' = -\frac{e^t}{2} \int_t^\infty e^{-t_1} f(t_1)\, dt_1 - c_2 e^{-t} + \frac{e^{-t}}{2} \int_0^t e^{t_1} f(t_1)\, dt_1$$

So far we have not used any particular norm. We now use the norm (1a). Then

$$(7) \quad \|u'\| \le \frac{1}{2}\|f\| + |c_2| + \frac{\|f\|}{2} = \|f\| + |c_2| \le \|u''\| + \|u\| + |c_2|$$

From this it is clear that, if $\|u\|$ and $\|u''\|$ are bounded, so also is $\|u'\|$. Furthermore, it follows from (6) that, if $|u|$ and $|u''| \to 0$ as $t \to \infty$, then so also does $|u'|$.

A similar result may be obtained using the norm (1b). This requires a bit more manipulation and is left as an exercise.   More precise reasoning yields the inequality

$$\|u'\|^2 \leq 4\|u\|\|u''\|$$ \hfill (8)

for both norms.

Exercise

Derive inequalities connecting $u^{(k)}$, $u^{(l)}$, and u for $k > l > 1$.

---

10. Oscillatory Equations.   In this section we consider the question of determining when all solutions of

$$u'' + \phi(t)u = 0 \tag{1}$$

have an infinite number of zeros in the interval $[0, \infty]$.   Equations whose solutions possess this property are called *oscillatory*, and the solutions are also named oscillatory.

Taking our cue from the equation

$$u'' + m^2 u = 0 \tag{2}$$

we prove first

Theorem 8.   *If all solutions of*

$$u'' + \phi(t)u = 0 \tag{3}$$

are oscillatory and if

$$\psi(t) \geq \phi(t) \tag{4}$$

then all solutions of

$$v'' + \psi(t)v = 0 \tag{5}$$

are oscillatory.

Furthermore, the following converse holds: If $\psi(t) \geq \phi(t)$ and some solutions of (5) are nonoscillatory, then some solutions of (3) must be nonoscillatory.

Proof.   We have

$$uv'' - vu'' + (\psi(t) - \phi(t))uv = 0 \tag{6}$$

Let t_1 and t_2 be two successive zeros of u and assume that $u \geq 0$ between t_1 and t_2.   Integrate over $[t_1, t_2]$, obtaining, since $uv'' - vu''$ is a perfect derivative,

$$(uv' - vu') \Big|_{t_1}^{t_2} + \int_{t_1}^{t_2} (\psi(t) - \phi(t))uv \, dt = 0 \tag{7}$$

and hence

$$(8) \qquad u'(t_1)v(t_1) - u'(t_2)v(t_2) + \int_{t_1}^{t_2} (\psi(t) - \phi(t))uv \, dt = 0$$

with $u'(t_1) > 0$ and $u'(t_2) < 0$. Consequently, it is impossible that any solutions of (5) be positive in the interval $[t_1, t_2]$, an even stronger result than stated.

To carry over this result to the equation

$$(9) \qquad \frac{d}{dt}\left(k_1(t)\frac{du}{dt}\right) + \phi_1(t)u = 0$$

we require a more recondite identity, due to Picone, namely,

$$(10) \quad \frac{d}{dt}\left[\frac{u}{v}\left(k_1 u'v - k_2 uv'\right)\right] = (\phi_1 - \phi_2)u^2 + (k_1 - k_2)u'^2 \\ + k_2\left(u' - \frac{uv'}{v}\right)^2$$

where v is the solution of the related equation with k_2 and ϕ_2. From this we easily obtain

Theorem 9. *If all solutions of*

$$(11) \qquad \frac{d}{dt}\left(k_1(t)\frac{du}{dt}\right) + \phi_1(t)u = 0$$

are oscillatory as $t \to \infty$ *and if*

$$(12) \qquad \begin{aligned} \phi_2 &\geq \phi_1 \\ k_2 &\geq k_1 > 0 \end{aligned}$$

then all solutions of

$$(13) \qquad \frac{d}{dt}\left(k_2(t)\frac{du}{dt}\right) + \phi_2(t)u = 0$$

are oscillatory.

Fortified by these comparison theorems, so useful in the Sturm-Liouville theory, we now seek some simple equations which are oscillatory. The simplest is (2), from which we conclude that, if

$$(14) \qquad \phi(t) \geq m^2 > 0$$

then all solutions of (3) are oscillatory. We shall give another proof below, which contains a method that can be applied to more general situations.

By means of a repeated change of variable, we can obtain some non-trivial comparison equations from (2). We shall demonstrate

Theorem 10. *If*

$$(15) \qquad \phi(t) \geq (1 + \epsilon) \frac{1}{4t^2}$$

or if

$$\phi(t) \geq \frac{1}{4t^2} + \frac{1}{4t^2 \log^2 t} + \cdots + (1 + \epsilon) \frac{1}{4t^2 \log^2 t \cdots \log_r^2 t}$$

where $r = 1, 2, \ldots$, *for* $t \geq t_0$ *and* $\epsilon > 0$, *then all solutions of* (3) *are oscillatory.*

Here

$$
(16) \qquad
\begin{aligned}
\log_1 (t) &= \log t \\
\log_2 (t) &= \log (\log t) \\
\log_r (t) &= \log (\log_{r-1} t)
\end{aligned}
$$

Proof. The proof follows by applying the substitutions

$$(17) \qquad t = e^{t_1}, \qquad t_1 = e^{t_2}, \qquad \ldots, \qquad t_n = e^{t_{n-1}}$$

in turn to (2). We know that the solutions of (2) are oscillatory only when $m^2 > 0$. Letting $t = e^{t_1}$, we obtain

$$(18) \qquad t_1^2 \frac{d^2 u}{dt_1^2} + t_1 \frac{du}{dt_1} + m^2 u = 0$$

Eliminating the term in du/dt by means of the substitution

$$(19) \qquad u = \frac{v}{\sqrt{t_1}}$$

we obtain

$$(20) \qquad \frac{d^2 v}{dt_1^2} + \frac{m^2 + \frac{1}{4}}{t_1^2} v = 0$$

Therefore all the solutions of (20) are oscillatory, if $m^2 > 0$.

Let us now make the substitution $t_2 = e^{t_1}$. The new equation is

$$(21) \qquad t_2^2 \frac{d^2 v}{dt_2^2} + t_2 \frac{dv}{dt_2} + \frac{m^2 + \frac{1}{4}}{\log^2 t_2} v = 0$$

Eliminating the term in dv/dt as above, we have

$$(22) \qquad \frac{d^2 w}{dt_2^2} + w \left(\frac{1}{4t_2^2} + \frac{m^2 + \frac{1}{4}}{t_2^2 \log^2 t_2} \right) w = 0$$

all of whose solutions are oscillatory. Continuing in this way, we obtain the result stated above.

The results are, in a sense, best possible, since each of the equations

$$(23) \qquad u'' + \frac{1}{4t^2} u = 0$$

$$u'' + \left(\frac{1}{4t^2} + \frac{1}{4t^2 \log^2 t} \right) u = 0$$

and so on, is nonoscillatory, and actually, as we see by retracing our steps to (2) with $m^2 = 0$, all solutions are eventually monotone.

If

$$(24) \qquad \int^{\infty} |g(t)| \frac{dt}{t} < \infty$$

then we know from Theorem 5 that all solutions of

$$(25) \qquad u'' + \frac{g(t)}{t^2} u = 0$$

are nonoscillatory. Once again, then, we see that Theorem 2 is close to a best possible result.

Let us now give another proof of the important and useful, albeit simple, result that (14) is a sufficient condition for oscillatory solutions. Let us assume that there exists a solution $u > 0$ for $t \geq t_0$. Then

$$(26) \qquad u'' = -\phi(t)u < 0$$

Hence u' is steadily decreasing. There are now two possibilities: $u' > 0$ for all $t \geq t_1$, or $u' < 0$ for $t \geq t_1$. Consider the first possibility. If $u' > 0$, u is monotone increasing, and $u \geq c_1$ for $t \geq t_1$. Thus

$$(27) \qquad u'' = -\phi(t)u \leq -a^2 c_1$$

whence, integrating, $u' \leq -a^2 t c_1 + c_2 \to -\infty$ as $t \to \infty$, which contradicts $u' \geq 0$. We are left with the second possibility, $u' < 0$ for $t \geq t_1$. Since u' is decreasing and is less than zero for $t \geq t_1$, $u' \leq -c_4$ for $t \geq t_1$; hence $u \leq -c_4 t + c_5$, a contradiction, as $t \to \infty$.

Returning to (27), we see that we may replace (14) by the weaker condition

$$(28) \qquad \phi(t) \geq 0, \qquad \int^{\infty} \phi(t) \, dt = \infty$$

Theorems on oscillation may also be derived by using the Riccati equation. As in (15) of Sec. 3, let $u'/u = v$, obtaining

$$(29) \qquad v' + v^2 + \phi(t) = 0$$

This equation is valid in an interval where $u \neq 0$. If $\phi(t) \geq 0$, we see that u'/u is monotone decreasing, in any interval in which it is continuous, and this property of the solutions of (3) is valid whether the solutions are oscillatory or not. Let $v = -w$, so that

(30) $$w' = w^2 + \phi(t)$$

If $\phi(t) \geq a^2 \geq 0$, we have

(31) $$w' \geq w^2 + a^2$$

which shows, by comparison with $w' = w^2 + a^2$, that $w \to \infty$ at a finite value of t and furthermore that every solution of (3) has a zero between two consecutive zeros of a solution of $u'' + a^2u = 0$; in other words, in every interval of length, $2\pi/a$.

11. $u'' + \phi(t)u = 0$, $\phi(t)$ **Periodic of Period π.** We now turn to the important and difficult question of deciding the boundedness of the solutions of the equation

(1) $$u'' + \phi(t)u = 0$$

where $\phi(t)$ is a continuous periodic function of period π.

The most important example of an equation of this type is the equation of Mathieu,

(2) $$u'' + (a + b \cos 2t)u = 0$$

which occurs in several important investigations of mathematical physics. The closely related equation

(3) $$u'' + \Big[\sum_{n=0}^{\infty} (a_n \cos nt + b_n \sin nt) \Big] u = 0$$

was encountered by Hill in his treatment of the motion of the moon. We shall not attempt to discuss these equations here, since (2) easily merits a separate treatise. The whole problem is one of great difficulty, and we shall content ourselves with proving an important general result due to Liapounoff.

Although we know, from the general representation theorem 11 of Sec. 15, Chap. 1, that every solution of (1) has the form

(4) $$u = e^{\rho t}p_1(t) + e^{\sigma t}p_2(t)$$

where $p_1(t)$ and $p_2(t)$ are periodic of period π, there exists no simple method for obtaining the constants ρ and σ explicitly, given $\phi(t)$.

There does exist, however, a simple criterion for boundedness:

Theorem 11. *If $\phi(t)$ is continuous, of period π, and if*

(5)   (a) $\displaystyle\int_0^{\pi} \phi(t)\, dt \geq 0$, $\phi(t) \not\equiv 0$

     (b) $\displaystyle\int_0^{\pi} |\phi(t)|\, dt \leq 4/\pi$

then all solutions of (1) are bounded as $t \to \pm \infty$.

Proof. Let u_1 and u_2 be a fundamental system of solutions of (1), that is, $u_1(0) = 1$, $u_1'(0) = 0$, $u_2(0) = 0$, $u_2'(0) = 1$. Since $u_1(t + \pi)$, and $u_2(t + \pi)$ are again solutions of (1), we must have

$$
(6) \qquad
\begin{aligned}
u_1(t + \pi) &= u_1(\pi)u_1(t) + u_1'(\pi)u_2(t) \\
u_2(t + \pi) &= u_2(\pi)u_1(t) + u_2'(\pi)u_2(t)
\end{aligned}
$$

Let λ_1 and λ_2 be the characteristic roots of the matrix

$$
(7) \qquad U = \begin{pmatrix} u_1(\pi) & u_1'(\pi) \\ u_2(\pi) & u_2'(\pi) \end{pmatrix}
$$

Applying (6) repeatedly we see that

$$
(8) \qquad \begin{pmatrix} u_1(t + n\pi) \\ u_2(t + n\pi) \end{pmatrix} = U^n \begin{pmatrix} u_1(\pi) \\ u_2(\pi) \end{pmatrix}
$$

Let us furthermore note that the determinant of U is equal to the Wronskian of u_1 and u_2 at $t = \pi$. Since the Wronskian is a constant for this equation, its value must be 1, the value determined at $t = 0$.

It follows from this that if (1) is to have unbounded solutions, U must have either distinct characteristic roots which are real, or multiple roots which are then either 1, 1 or -1, -1. If U has complex roots, which are necessarily distinct and of absolute value 1, the matrices U^n will be uniformly bounded as $n \to \pm \infty$, implying the boundedness of the solutions of (1). Hence if unbounded solutions exist, we must have either

$$
(9) \qquad U = T^{-1} \begin{pmatrix} \lambda_1 & 0 \\ 0 & \lambda_2 \end{pmatrix} T \qquad \text{or} \qquad U = T^{-1} \begin{pmatrix} \lambda & 1 \\ 0 & \lambda \end{pmatrix} T
$$

where λ_1, λ_2 are real and $\lambda = \pm 1$ in the second representation. Returning to (6) this implies

$$
(10) \qquad T \begin{pmatrix} u_1(t + \pi) \\ u_2(t + \pi) \end{pmatrix} = \begin{pmatrix} \lambda_1 & 0 \\ 0 & \lambda_2 \end{pmatrix} T \begin{pmatrix} u_1(t) \\ u_2(t) \end{pmatrix}
$$

or

$$
(10') \qquad T \begin{pmatrix} u_1(t + \pi) \\ u_2(t + \pi) \end{pmatrix} = \begin{pmatrix} \lambda & 1 \\ 0 & \lambda \end{pmatrix} T \begin{pmatrix} u_1(t) \\ u_2(t) \end{pmatrix}
$$

and, in consequence, the existence of a real, nontrivial solution satisfying

$$
(11) \qquad u_3(t + \pi) = \lambda_1 u_3(t)
$$

We now derive a contradiction from this fact. It follows from (11) that $u_3(t)$ either is never zero or else is infinitely often zero. Let us suppose that $u_3(t)$ is never zero. Then, from (1),

$$
(12) \qquad \int_0^\pi \frac{u_3''}{u_3} \, dt + \int_0^\pi \phi(t) \, dt = 0
$$

Integrating by parts, we obtain

$$(13) \qquad \frac{u_3'}{u_3}\Big|_0^\pi + \int_0^\pi \frac{(u_3')^2}{u_3^2}\, dt + \int_0^\pi \phi(t)\, dt = 0$$

Since $u_3'(\pi)/u_3(\pi) = u_3'(0)/u_3(0)$, the above result contradicts (5a).

Now consider the case where u_3 has zeros. The distance between adjacent zeros is always less than or equal to π, from (11). We now derive a general inequality, from which we shall derive a contradiction to (5b).

Lemma 9. *Let* $u(a) = u(b) = 0$, *where* $0 < b - a \le \pi$ *and* $u(t) > 0$ *in* (a,b). *Then*

$$(14) \qquad \int_a^b \left| \frac{u''}{u} \right| dt > \frac{4}{b-a}$$

Proof of Lemma

$$(15) \qquad \int_a^b \left| \frac{u''}{u} \right| dt > (u_{\max})^{-1} \int_a^b |u''|\, dt > (u_{\max})^{-1} \max_{a \le t_1 < t_2 \le b} |u'(t_2) - u'(t_1)|$$

Let $u_{\max} = u(a + l_1) = u(b - l_2)$, $l_1 + l_2 = b - a$. Then by Rolle's theorem,

$$(16) \qquad u'(t_2) = l_1^{-1}u_{\max}, \qquad -u'(t_1) = l_2^{-1}u_{\max}$$

for some t_1, t_2, where $a < t_2 < a + l_1 = b - l_2 < t_1 < b$. Hence from (15)

$$(17) \qquad \int_a^b \left| \frac{u''}{u} \right| dt > l_1^{-1} + l_2^{-1} = \frac{l_1 + l_2}{l_1 l_2} > \frac{4}{b-a}$$

since $(l_1 + l_2)/2 \ge \sqrt{l_1 l_2}$.

Now since u_3 satisfies (1), with $0 < b - a \le \pi$, we have

$$(18) \qquad \frac{4}{\pi} < \int_a^b \left| \frac{u_3''}{u_3} \right| dt \le \int_0^\pi \left| \frac{u_3''}{u_3} \right| dt = \int_0^\pi |\phi(t)|\, dt \le \frac{4}{\pi}$$

a contradiction.

12. The Asymptotic Behavior of the Solutions of the Equation $u'' - (1 + \phi(t))u = 0$. **Résumé of Results.** Let us begin with the equation

$$(1) \qquad u'' - (1 + \phi(t))u = 0$$

whose theory is simpler than that of $u'' + (1 + \phi(t))u = 0$ and, in consequence, more complete. Under the assumption that $\phi(t) \to 0$ as $t \to \infty$, we know of the existence of two solutions u_1 and u_2, for which

$$(2) \qquad \frac{u_1'}{u_1} \to 1, \qquad \frac{u_2'}{u_2} \to -1$$

Theorem 7 of Chap. 2 shows that there exist two solutions u_1 and u_2 such that

$$(3) \quad \exp\left[t - c_1 \int_0^t |\phi(t)|\, dt\right] \leq u_1 \leq \exp\left[t + c_1 \int_0^t |\phi(t)|\, dt\right]$$

$$\exp\left[-t - c_1 \int_0^t |\phi(t)|\, dt\right] \leq u_2 \leq \exp\left[-t + c_1 \int_0^t |\phi(t)|\, dt\right]$$

Exercise

Do solutions satisfying (3) automatically satisfy (2)?

If we now assume that $\int^\infty |\phi(t)|\, dt < \infty$, we can assert the existence of two solutions u_1 and u_2 for which, as $t \to \infty$,

$$(4) \qquad\qquad u_1 \sim e^t$$
$$u_2 \sim e^{-t}$$

as follows from Theorems 7 and 8 of Chap. 2.

The result of (4) can be considerably improved if we assume correspondingly more about $\phi(t)$. Thus, if $\phi(t)$ possesses an asymptotic expansion

$$(5) \qquad\qquad \phi(t) \sim \frac{c_2}{t^2} + \frac{c_3}{t^3} + \cdots + \frac{c_n}{t^n} + \cdots$$

then there are two solutions u_1 and u_2 satisfying

$$(6) \qquad\qquad u_1 \sim e^t \sum_{k=0}^\infty a_k t^{-k}$$

$$u_2 \sim e^{-t} \sum_{k=0}^\infty b_k t^{-k}$$

as $t \to \infty$.

It follows from the general result of Theorem 8 of Chap. 2 that, if

$$(7) \qquad\qquad \int^\infty |\phi'(t)|\, dt < \infty$$

there are two solutions u_1 and u_2 for which

$$(8) \qquad\qquad u_1 \sim \exp \int_0^t \sqrt{1 + \phi(t)}\, dt$$

$$u_2 \sim \exp\left[-\int_0^t \sqrt{1 + \phi(t)}\, dt\right]$$

as $t \to \infty$.

From this it follows that, if we assume in addition that

$$\int^\infty \phi^2(t)\, dt < \infty$$

then there exist two solutions u_1 and u_2 such that

$$
(9) \qquad u_1 \sim \exp\left[t + \tfrac{1}{2} \int_0^t \phi(t_1)\, dt_1 \right]
$$
$$
u_2 \sim \exp\left[-t - \tfrac{1}{2} \int_0^t \phi(t_1)\, dt_1 \right]
$$

All the above theorems are special cases of more general theorems valid for systems. Let us now turn to some methods particularly applicable to second-order equations.

13. The Equation $u'' - (1 + \phi(t))u = 0$, where $\phi(t) \to 0$ as $t \to \infty$. We have observed in the above résumé of results that, if $\phi(t) \to 0$ as $t \to \infty$, there exist two solutions u_1 and u_2 for which

$$
(1) \qquad \frac{u_1'}{u_1} \to 1, \qquad \frac{u_2'}{u_2} \to -1
$$

We propose to give another proof of this result, employing a very useful and interesting method. First we shall show that a solution exists for which $u_1'/u_1 \to 1$, and then we shall use the result of Lemma 3, stating that

$$
(2) \qquad u_2 = u_1 \int_t^\infty \frac{dt}{u_1^2}
$$

is another solution, to obtain a solution satisfying the second condition in (1).

Choose t_0 large enough so that $1 + \phi(t) \geq 1 - \epsilon$ for $t \geq t_0$. Choose u as the solution of

$$
(3) \qquad u'' - (1 + \phi(t))u = 0
$$

for which $u'(t_0) = 2$ and $u(t_0) = 1$, and set $v = u'/u$, so that v satisfies the Riccati equation

$$
(4) \qquad v' + v^2 - (1 + \phi(t)) = 0, \qquad v(t_0) = 2
$$

Consider Fig. 1. From the differential equation (3), we see that v' is negative at t_0, whence v decreases until v hits the curve $w = \sqrt{1 + \phi(t)}$. At the point of intersection P, v has a minimum, and v begins to increase until it hits the curve again. At the next point of intersection, v has a maximum and turns downward again, and so on. Since the maxima and minima of v are contained between the maxima and minima of $w = \sqrt{1 + \phi(t)}$, it is clear that $v \to 1$ as $t \to \infty$. We have con-

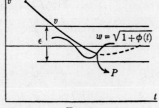

FIG. 1.

sidered the case where $1 + \phi(t)$ is oscillating as $t \to \infty$. The other case is even simpler.

Since $v = u'/u$, we have proved the existence of a solution u, for which $u'/u \to 1$. Let us now use this solution to derive another solution for which $u'/u \to -1$. Since $u \geq e^{(1-\epsilon)t}$ for $t \geq t_0$, the function

$$(5) \qquad w = u \int_t^\infty \frac{dt}{u^2}$$

exists and, by virtue of Lemma 3, is another solution of (1), as noted above.

From this, we obtain

$$(6) \qquad \frac{w'}{w} = \frac{u'\left(\int_t^\infty dt/u^2\right) - (1/u)}{u \int_t^\infty (dt/u^2)} = \frac{u'}{u} - \frac{1/u^2}{\int_t^\infty (dt/u^2)}$$

We now require the following lemma:

Lemma 10. *Let*

$$(7) \qquad \lim_{t \to \infty} \frac{f'(t)}{g'(t)} = a, \qquad f(t), g(t) \to 0 \text{ as } t \to \infty$$

g' being one-signed for $t \geq t_0$. Then

$$(8) \qquad \lim_{t \to \infty} \frac{f(t)}{g(t)} = a$$

Proof. Take $g'(t)$ as positive without loss of generality. Then for $t \geq t_0$,

$$(9) \qquad a - \epsilon \leq \frac{f'(t)}{g'(t)} \leq a + \epsilon$$

or

$$(10) \qquad (a - \epsilon)g'(t) \leq f'(t) \leq (a + \epsilon)g'(t)$$

Integrating between t and ∞, we obtain, for $t \geq t_0$,

$$(11) \qquad (a + \epsilon)g(t) \leq f(t) \leq (a - \epsilon)g(t)$$

Since this is true for any $\epsilon > 0$, with t_0 sufficiently large, (8) holds. Applying this to the second term on the right-hand side of (6), we obtain

$$(12) \qquad \lim_{t \to \infty} \frac{1/u^2}{\int_t^\infty dt/u^2} = \lim_{t \to \infty} \frac{-2u'/u^3}{-1/u^2} = 2 \lim_{t \to \infty} \frac{u'}{u} = 2$$

Therefore, $w'/w \to -1$ as $t \to \infty$.

Exercise

Show that $v = 1$ is a stable limit of solutions of (4) and that $v = -1$ is an unstable limit.

---

14. The Liouville Transformation. In a previous section, Lemma 5 of Sec. 3, we have shown how the Liouville transformation may be used to transform the equation

$$(1) \qquad\qquad u'' \pm \phi^2(t)u = 0$$

into the form

$$(2) \qquad\qquad u'' \pm (1 + \phi_1(t))u = 0$$

where $\phi_1(t)$ in general is a function which approaches zero as $t \to \infty$, if $\phi(t) \to \infty$.

We now wish to point out that a repetition of this transformation will frequently transform (2) into an equation of the same form amenable to the results of Sec. 12. Making the change of variable

$$(3) \qquad\qquad s = \int_0^t \sqrt{1 + \phi(t)}\, dt$$

(set $\phi_1 \equiv \phi$), (2) is transformed into

$$(4) \qquad\qquad \frac{d^2u}{ds^2} + \frac{\phi'(t)}{2(1 + \phi)^{3/2}} \frac{du}{ds} \pm u = 0$$

The further change of variable

$$(5) \qquad\qquad u = v \exp\left[-\tfrac{1}{2} \int_0^s a(s)\, ds \right]$$

where

$$(6) \qquad\qquad a(s) = \frac{\phi'(t)}{2(1 + \phi)^{3/2}}$$

yields the equation

$$(7) \qquad\qquad v''(s) + \left[\pm 1 - \frac{1}{2}\frac{da(s)}{ds} - \frac{a^2(s)}{4} \right] v = 0$$

which is again of the same form as (2).

This method is particularly applicable if $\phi(t) = 1/t^a$, $1/\log t$, and generally any rational function in t, e^t, $\log t$, and so on. Sometimes, several successive transformations are required to meet the conditions of Sec. 12.

Exercises

1. Find the asymptotic form of the solutions of $u'' - e^t u = 0$,

$$u'' - t^n u = 0$$

$u'' - (\log \log t)u = 0$.

2. Formulate a general criterion in terms of ϕ, ϕ', and ϕ'' which may be used to determine the asymptotic behavior of the solutions of

$$u'' = \phi^2(t)u$$

Exhibit some equations which escape this criterion.

────────────

15. The Equation $u'' - (1 + \phi(t))u = 0$, $\int^\infty |\phi'(t)|\,dt = \infty, \int^\infty \phi^2(t)\,dt$ $< \infty$.   There is an extensive class of equations of the form (2) of Sec. 14 which escape the previous analysis.   A simple example is

$$(1) \qquad u'' - \left(1 + \frac{\sin t}{t^a}\right)u = 0, \qquad \frac{1}{2} < a \le 1$$

Here

$$(2) \qquad \int^\infty \left|\frac{\sin t}{t^a}\right|\,dt = \infty, \qquad \int^\infty \left|\frac{d}{dt}\frac{\sin t}{t^a}\right|\,dt = \infty$$

Let us now present a method based upon the connection between the equation

$$(3) \qquad u'' - (1 + \phi(t))u = 0$$

and the Riccati equation

$$(4) \qquad v' + v^2 - (1 + \phi(t)) = 0, \qquad v = \frac{u'}{u}$$

Setting $v = 1 + w$, we obtain for w the equation

$$(5) \qquad w' = -2w - w^2 + \phi(t)$$

We know that there is a solution which approaches zero as $t \to \infty$.   Let us obtain an upper bound for this solution in terms of $\phi(t)$.   We have

$$(6) \qquad w = \int_0^t e^{-2(t-t_1)}\phi(t_1)\,dt_1 - \int_0^t e^{-2(t-t_1)}w^2(t_1)\,dt_1$$

where for convenience of notation we represent the lower limit as zero. It is no loss of generality to assume that $|\phi(t)| \le \epsilon$ for $t \ge 0$.   Using the method of successive approximations, we set

(7)
$$w_0 = \int_0^t e^{-2(t-t_1)}\phi(t_1)\,dt_1$$

$$w_{n+1} = \int_0^t e^{-2(t-t_1)}\phi(t_1)\,dt_1 - \int_0^t e^{-2(t-t_1)}w_n^2(t_1)\,dt_1$$

Employing the bound for $\phi(t)$, we have $|w_0| \leq \epsilon/2$. Hence, if $|w_n| \leq \epsilon$, we have

(8)
$$|w_{n+1}| \leq \int_0^t e^{-2(t-t_1)}|\phi(t_1)|\,dt_1 + \epsilon^2 \int_0^t e^{-2(t-t_1)}\,dt_1$$

$$\leq \frac{\epsilon}{2} + \frac{\epsilon^2}{2} \leq \epsilon$$

for $\epsilon \leq 1$. The inequality thus holds for all n.

From (7) we obtain

(9)
$$|w_{n+1}| \leq \int_0^t e^{-2(t-t_1)}|\phi(t_1)|\,dt_1 + \epsilon \int_0^t e^{-2(t-t_1)}|w_n|\,dt_1$$

Let us show, inductively, that $|w_n| \leq 2\int_0^t e^{-(t-t_1)}|\phi(t_1)|\,dt_1$, if ϵ is sufficiently small. The inequality is certainly true for $n = 0$.

From (9) we obtain

(10)
$$|w_{n+1}| \leq \int_0^t e^{-(t-t_1)}|\phi(t_1)|\,dt_1 + 2\epsilon \int_0^t e^{-2(t-t_1)}$$
$$\left(\int_0^{t_1} e^{-2(t_1-t_2)}|\phi(t_2)|\,dt_2 \right)dt_1$$

$$\leq \int_0^t e^{-(t-t_1)}|\phi(t_1)|\,dt_1 + 2\epsilon \int_0^t e^{-2t} \left(\int_0^{t_1} e^{2t_2}|\phi(t_2)|\,dt_2 \right)dt_1$$

The second term on the right is

(11)
$$2\epsilon e^{-2t}\int_0^t \left[\int_0^{t_1} e^{2t_2}|\phi(t_2)|\,dt_2 \right]dt_1 = 2\epsilon e^{-2t}\int_0^t (t - t_2)e^{2t_2}|\phi(t_2)|\,dt_2$$

$$\leq 2\epsilon \int_0^t e^{-(t-t_2)}|\phi(t_2)|\,dt_2$$

Consequently,

(12)
$$|w_{n+1}| \leq \int_0^t e^{-(t-t_1)}|\phi(t_1)|\,dt_1 + 2\epsilon \int_0^t e^{-(t-t_2)}|\phi(t_2)|\,dt_2$$

$$< 2\int_0^t e^{-(t-t_1)}|\phi(t_1)|\,dt_1$$

if $\epsilon \leq \frac{1}{2}$.

It is easy to show, using the methods we have developed previously, that w_n converges to w, a solution of (6), which satisfies the inequality

(13)
$$|w| \leq 2\int_0^t e^{-(t-t_1)}|\phi(t_1)|\,dt_1$$

Using Schwarz's inequality, we obtain

(14)
$$w^2 \leq 2\left(\int_0^t e^{-(t-t_1)}\,dt_1 \right)\left(\int_0^t e^{-(t-t_1)}\phi^2(t_1)\,dt_1 \right)$$

$$\leq 2\int_0^t e^{-(t-t_1)}\phi^2(t_1)\,dt_1$$

Returning to (6), we obtain

$$(15) \quad w = \int_0^t e^{-2(t-t_1)} \phi(t_1) \, dt_1 + O\left(\int_0^t e^{-2(t-t_1)} \left[\int_0^{t_1} e^{-(t_1-t_2)} \phi^2(t_2) \, dt_2\right] dt_1\right)$$
$$= \int_0^t e^{-2(t-t_1)} \phi(t_1) \, dt_1 + O\left(\int_0^t e^{-(t-t_1)} \phi^2(t_1) \, dt_1\right)$$

Since $u'/u = 1 + w$, we see that it is the behavior of $\int_0^t w \, dt$ which determines the asymptotic behavior of u. Using (15), we have

$$(16) \quad \int_0^t w \, dt_1 = \int_0^t \left[\int_0^{t_1} e^{-2(t_1-t_2)} \phi(t_2) \, dt_2\right] dt_1$$
$$+ O\left(\int_0^t \left[\int_0^{t_1} e^{-(t_1-t_2)} \phi^2(t_2) \, dt_2\right] dt_1\right)$$

Interchanging the order of integration in both integrals, we obtain

$$(17) \quad \int_0^t w \, dt_1 = \tfrac{1}{2} \int_0^t \phi(t_2) \, dt_2 - \tfrac{1}{2} \int_0^t e^{-2(t-t_2)} \phi(t_2) \, dt_2$$
$$+ O\left(\int_0^t \phi^2(t_2) \, dt_2\right)$$

Since $\phi(t) \to 0$ as $t \to \infty$ and $\int_0^\infty \phi^2(t) \, dt < \infty$, we see that

$$(18) \qquad u = \exp\left(t + \tfrac{1}{2} \int_0^t \phi(t_1) \, dt_1 + c_1 + o(1)\right)$$

as $t \to \infty$.

The asymptotic behavior of a second linearly independent solution may be determined, as usual, by using the second solution $u_2 = u \int_t^\infty dt/u^2$. Alternatively, one may consider the equation derived by setting $v = -1 + w$,

$$(19) \qquad w' = 2w - w^2 + \phi(t)$$

The corresponding integral equation for solutions tending to zero is

$$(20) \qquad w = \int_t^\infty e^{2(t-t_1)} \phi(t_1) \, dt_1 - \int_t^\infty e^{2(t-t_1)} w^2(t_1) \, dt_1$$

The final result is

Theorem 12. *If*

$$(21) \quad (a) \ \ \phi(t) \to 0 \ as \ t \to \infty$$
$$(b) \ \ \int_0^\infty \phi^2(t) \, dt < \infty$$

there are two solutions of $u'' - (1 + \phi(t))u = 0$ possessing the asymptotic forms

(22)
$$u_1 = \exp\left(t + \tfrac{1}{2}\int_0^t \phi(t)\,dt + o(1)\right)$$
$$u_2 = \exp\left(-t - \tfrac{1}{2}\int_0^t \phi(t)\,dt + o(1)\right)$$

as $t \to \infty$.

Exercise

Use (20) to derive the asymptotic form of u_2.

---

16. The Equation $u'' - (1 + \phi(t))u = 0$, $\int^\infty |\phi(t)|^n\,dt < \infty$. The method employed in the preceding section is applicable to any equation of the form

(1)
$$u'' - (1 + \phi(t))u = 0$$

where $\phi \to 0$ as $t \to \infty$ and $\int^\infty |\phi|^n\,dt < \infty$ for some $n > 0$. In practice, however, the algebraic complexities become overwhelming. We shall content ourselves with stating the result for $n = 3$ and leaving the proof as an exercise.

Theorem 13. *If*

(2) (a) $\phi(t) \to 0$ *as* $t \to \infty$
   (b) $\int^\infty |\phi(t)|^3\,dt < \infty$

there are two solutions of

(3)
$$u'' - (1 + \phi(t))u = 0$$

possessing the asymptotic expansions

(4) $u_1 = \exp\Big(t + \tfrac{1}{2}\int_0^t \phi(t_1)\,dt_1$
$$- \tfrac{1}{2}\int_0^t \phi(t_2)\int_0^{t_2} \phi(t_3)e^{-2t_3+2t_2}\,dt_3\,dt_2 + o(1)\Big)$$

   $u_2 = \exp\Big(-t - \tfrac{1}{2}\int_0^t \phi(t_1)\,dt_1$
$$+ \tfrac{1}{2}\int_0^t \phi(t_2)\int_0^{t_2} \phi(t_3)e^{-2t_3+2t_2}\,dt_3\,dt_2 + o(1)\Big)$$

It appears to be impossible to transform the double integral into any simpler form involving single integrals. We would expect this term to behave like $\int_0^t \phi^2(t_1)\,dt_1/8$, by analogy with the expansion

(5)
$$\int_0^t \sqrt{1 + \phi(t_1)}\,dt_1 = t + \tfrac{1}{2}\int_0^t \phi(t_1) - \tfrac{1}{8}\int_0^t \phi^2(t_1)\,dt_1 + \cdots$$

In any specific case, such as $\phi = (\sin t)/\sqrt{t}$, integration by parts is readily applicable.

17. Extension to Higher-order Equations. The essence of the above method lies in the fact that the substitution

$$(1) \qquad\qquad v = \frac{u'}{u}$$

transforms the linear equation

$$(2) \qquad\qquad u^{(n)} + a_1(t)u^{(n-1)} + \cdots + a_n(t)u = 0$$

into a nonlinear equation of the $(n - 1)$st order, of Poincaré-Liapounoff type.

As an example of how the method might be applied further, let us consider the equation

$$(3) \qquad\qquad u^{(3)} + a_1(t)u^{(2)} + a_2(t)u^{(1)} + a_3(t)u = 0$$

where $a_i(t) \to a_i$ as $t \to \infty$. Setting $v = u'/u$, the equation for v is

$$(4) \qquad v'' + 3vv' + v^3 + a_1(t)(v' + v^2) + a_2(t)v + a_3(t) = 0$$

Let r_1 be a root of the algebraic equation

$$(5) \qquad\qquad r^3 + a_1r^2 + a_2r + a_3 = 0$$

and set $v = r_1 + w$. The equation for w is

$$(6) \quad w'' + 3(w + r_1)w' + r_1^3 + 3r_1^2w + a_1(t)(w' + r_1^2 + 2r_1w + w^2) \\ + 3r_1w^2 + w^3 + a_2(t)r_1 + a_2(t)w + a_3(t) = 0$$

The approximating linear equation is

$$(7) \quad w'' + 3r_1w' + 3r_1^2w + a_1(t)w' + 2r_1a_1(t)w + a_2(t)w + r_1^3 + a_1(t)r_1^2 \\ + a_2(t)r_1 + a_3(t) = 0$$

The asymptotic behavior of the solution is determined by the algebraic nature of the roots of

$$(8) \qquad\qquad \alpha^2 + (3r_1 + a_1)\alpha + (3r_1^2 + 2r_1a_1 + a_2) = 0$$

If the roots of this equation are distinct, with nonzero real part, the familiar method of successive approximations may be used to obtain the solutions of (6) and to ascertain their asymptotic behavior, provided that various integrability conditions are satisfied, for example,

$$(9) \qquad\qquad \int^\infty |a_i - a_i(t)|^n \, dt < \infty$$

for some $n \geq 1$. If there are roots with zero real part, we must have recourse to conditions of the type of Theorem 10 of Chap. 2.

18. The Equation $u'' + (1 + \phi(t))u = 0$. Résumé of Results. We now consider the equation

$$(1) \qquad\qquad u'' + (1 + \phi(t))u = 0, \qquad \phi(t) \to 0 \text{ as } t \to \infty$$

whose theory is far more difficult because of the oscillatory behavior of the solutions. There is no analogue here of the theorems concerning the asymptotic behavior of u'/u. However, if $\int^{\infty} |\phi(t)| \, dt < \infty$, it follows from Theorem 10 of Chap. 2 that there are two solutions of (1), satisfying

$$
(2) \qquad
\begin{aligned}
u_1 &= (\sin t)(1 + o(1)) \\
u_2 &= (\cos t)(1 + o(1))
\end{aligned}
$$

as $t \to \infty$.

If we make the further assumption that $\phi(t)$ has an asymptotic series

$$
(3) \qquad \phi(t) \sim \frac{c_2}{t^2} + \frac{c_3}{t^3} + \cdots + \frac{c_n}{t^n} + \cdots
$$

then two solutions exist, u_1 and u_2, having, respectively, the forms, as $t \to \infty$,

$$
(4) \qquad u_1 = \sin t \left[\sum_{k=0}^{n} a_k t^{-k} + O(t^{-n-1}) \right]
$$

$$
u_2 = \cos t \left[\sum_{k=0}^{n} b_k t^{-k} + O(t^{-n-1}) \right]
$$

From Theorem 10 of Chap. 2, it follows that, if

$$
(5) \qquad \int^{\infty} |\phi'(t)| \, dt < \infty
$$

there are two solutions for which

$$
(6) \qquad
\begin{aligned}
u_1 &= \left[\exp\left(i \int_0^t \sqrt{1 + \phi(t_1)} \, dt_1 \right) \right] (1 + o(1)) \\
u_2 &= \left[\exp\left(-i \int_0^t \sqrt{1 + \phi(t_1)} \, dt_1 \right) \right] (1 + o(1))
\end{aligned}
$$

as $t \to \infty$. Consequently, if in addition

$$
(7) \qquad \int^{\infty} \phi^2(t) \, dt < \infty
$$

we can refine this to

$$
(8) \qquad
\begin{aligned}
u_1 &= \left[\exp\left(it + \frac{i}{2} \int_0^t \phi(t_1) \, dt_1 \right) \right] (1 + o(1)) \\
u_2 &= \left[\exp\left(-it - \frac{i}{2} \int_0^t \phi(t_1) \, dt_1 \right) \right] (1 + o(1))
\end{aligned}
$$

If both $\int^{\infty} |\phi(t)| \, dt$ and $\int^{\infty} |\phi'(t)| \, dt$ are infinite, the problem requires more delicacy. In the next section we consider problems of this type.

19. The Equation $u'' + (1 + \phi(t))u = 0$ **(Continuation).** In the previous sections we have discussed the asymptotic behavior of the solutions of $u'' + (1 + \phi(t))u = 0$ under the assumptions either that $\int^{\infty} |\phi(t)|\, dt < \infty$ and $\int^{\infty} |\phi'(t)|\, dt < \infty$ or that ϕ was a sum of functions satisfying these conditions. If we consider the equation

$$(1) \qquad u'' - \left(1 + \frac{\sin t}{t^\alpha}\right)u = 0, \qquad 0 < \alpha < 1$$

we see that these criteria fail and that more subtle techniques must be employed.

To extend the range of our previous methods, we proceed as follows: As before, the differential equation $u'' + (1 + \phi(t))u = 0$ is transformed into the integral equation

$$(2) \qquad u = v - \int_0^t \sin\,(t - t_1)\phi(t_1)u(t_1)\, dt_1$$

where $v = c_1 \cos t + c_2 \sin t$. We now iterate once, obtaining

$$(3) \quad u = v - \int_0^t \sin\,(t - t_1)\phi(t_1)\left[v(t_1) - \int_0^{t_1} \sin\,(t_1 - t_2)\phi(t_2)u(t_2)\, dt_2 \right] dt_1$$
$$= v - \int_0^t \sin\,(t - t_1)\phi(t_1)v(t_1)\, dt_1$$
$$+ \int_0^t \left[\sin\,(t - t_1)\phi(t_1) \int_0^{t_1} \sin\,(t_1 - t_2)\phi(t_2)u(t_2)\, dt_2 \right] dt_1$$

Inverting the order of integration in the second integral, we obtain

$$(4) \quad u = v - \int_0^t \sin\,(t - t_1)\phi(t_1)v(t_1)\, dt_1$$
$$+ \int_0^t \phi(t_1)\left[\int_{t_1}^t \sin\,(t - t_1)\sin\,(t_1 - t_2)\phi(t_2)\, dt_2 \right] u(t_1)\, dt_1$$

We now leave as an exercise the proof of the following theorem:

Theorem 14. *Sufficient conditions that all solutions of the equation* $u'' + (1 + \phi(t))u = 0$ *be bounded are that*

$(5) \quad (a)$ $\int_{t_0}^t \phi(t_1)\, dt_1$, $\int_{t_0}^t \phi(t_1) \sin 2t_1\, dt_1$, *and* $\int_{t_0}^t \phi(t_1) \cos 2t_1\, dt_1$ *be uniformly bounded for* $t_0 \leq t < \infty$, *and*

$\quad (b)$ $\int_{t_0}^t \left| \phi(t_1) \int_{t_1}^t \sin\,(t - t_1)\sin\,(t_1 - t_2)\phi(t_2)\, dt_2 \right| dt_1 < k < 1$ *for all* $t \geq t_0$ *for some* t_0.

Exercises

1. Prove that all solutions of $u'' + (1 + \sin at/t^b)u = 0$ are bounded if $a \neq 2$ and $b > \frac{1}{2}$.

2. Prove that $u'' + (1 + \sin 2t/t)u = 0$ has unbounded solutions, and determine their asymptotic behavior.

3. Prove that all solutions of $u'' + (1 + \sin t^{2a})u = 0$ are bounded if $a > 1$.

4. Prove the analogue of Theorem 14 for the general equation

$$u'' + (a(t) + \phi(t))u = 0$$

and for the vector-matrix equation $z' = (A + B(t))z$.

---

20. The Equations $\epsilon(t)u'' + u' + u = 0$ and $u'' + a(t)u' + u = 0$. Let us turn our attention to the two equations

$$(1) \qquad \epsilon(t)u'' + u' + u = 0$$

and

$$(2) \qquad u'' + a(t)u' + u = 0$$

where $\epsilon(t) \to 0$ as $t \to \infty$ and $a(t) \to \infty$ as $t \to \infty$.

Comparing (2) with the corresponding equation where $a(t)$ is a positive constant, we should expect that every solution of (2) would approach zero as $t \to \infty$, and the more rapidly the larger $a(t)$. We might also expect to find a solution of (1) asymptotic to e^{-t} as $t \to \infty$, if $\epsilon(t) \to 0$ as $t \to \infty$.

It is rather interesting to observe that the problems are equivalent, since the substitution $u = ve^{-t}$ transforms (1) into

$$(3) \qquad v'' + \left(\frac{1}{\epsilon(t)} - 2\right)v' + v = 0$$

Consequently, if every solution of (2) approaches zero as $t \to \infty$ whenever $a(t) \to \infty$ as $t \to \infty$, there are no solutions of (1) asymptotic to e^{-t}; and conversely, if there is always a solution of (1) which is asymptotic to e^{-t} as $t \to \infty$ for every $\epsilon(t)$ which approaches zero as $t \to \infty$, then all solutions of (2) cannot approach zero as $t \to \infty$. An interesting clash of intuitions!

Of course, using our intuition in a different way, we might suspect that (2), for large $a(t)$, would have solutions corresponding to the two approximations

$$(4) \qquad \begin{aligned} u'' + a(t)u' &= 0 \\ a(t)u' + u &= 0 \end{aligned}$$

Hence, if $a(t) \to \infty$ rapidly enough so that $\int^{\infty} dt/a(t) < \infty$, we should look for a bounded solution, not approaching zero.

The reader will find it interesting to apply the methods of the preceding sections, in particular the Liouville transformation, to (1) and (2) to determine conditions under which the various situations occur.

Exercises

1. Prove that, if $a(t) > 0$ for all $t \geq 0$, then any solution of

$$u'' + a(t)u = 0$$

satisfies the inequality

$$u^2 < \frac{c_1}{a(t)} \exp\left(\int_0^t \frac{|a'(t)|}{a(t)}\, dt\right)$$

for $t \geq 0$.

2. If $\int^{\infty} |a(t)|\, dt < \infty$, all solutions of $u'' + a(t)u = 0$ cannot be bounded.

3. All solutions of $d(\phi(t)du/dt)/dt + a(t)u = 0$ are bounded provided that $a(t) > 0$, $\phi(t) > 0$, $d(a(t)\phi(t))/dt > 0$, for $t > t_0$. (Butlewski.)

4. All solutions of $d(\phi(t)du/dt)/dt + \sum_{i=0}^{m} a_{2i+1}(t)u^{2i+1} = 0$ are bounded, provided that $\phi(t) > 0$, $a_{2i+1}(t) > 0$, $d(a_{2i+1}\phi)/dt > 0$ for $t \geq t_0$. (Butlewski.)

5. The equation $u'' - \phi(t)u = 0$ can have no nontrivial solution bounded for $-\infty < t < \infty$ if $\phi(t) > \alpha > 0$ for $-\infty < t < \infty$. (Murray.)

6. If $0 < b^2 < \phi(t) < a^2$ for $-\infty < t < \infty$ and if $|\psi(t)| < c_1$ for $-\infty < t < \infty$, there is one and only one solution of $u'' - \phi(t)u = \psi(t)$ which is bounded for $-\infty < t < \infty$. (Murray.)

7. Consider the equation $u'' + (a^2 + \phi(t))u = 0$. Let $a(t)$ be any monotone increasing function such that $a'(t) = O(1)$ as $t \to \infty$. Then there exists a $\phi(t)$ such that, for large t, $\int_0^t \phi(t)\, dt < a(t)$, and $\varlimsup_{t \to \infty}$ (log $|u|/a(t)) \geq 1/\pi$. (Levinson.)

8. If $|a(t)| < c_1$ for $t \geq t_0$, all solutions of $u'' + a(t)u = 0$ cannot belong to $L^2(0, \infty)$.

9. If $|a(t)| < c_1$, with $t > t_0$, then, if u belongs to $L^2(0, \infty)$, du/dt also belongs to $L^2(0, \infty)$. The result holds if we require only $a(t) < c_1$. (Wintner.)

10. Consider the equation $u'' = f(u,t)$, where $f(u,t)$ has the same sign as u and is continuous for all u for $t \geq t_0$, and where a solution is determined by the values of u and u' at any point in the t interval (t_0, ∞). Then only one of the functions u or u' can vanish for $t \geq t_0$ and only once. As $t \to \infty$, two cases are possible:

(a) $u \to \pm \infty$ monotonically

(b) u and $u' \to 0$, both monotonically, one increasing, the other decreasing

(Kneser.)

11. If $0 < c_1 < f(t) < c_2$, there is one and only one solution of $u'' = f(t)u$ which remains finite as $t \to \infty$, and this approaches zero as $t \to \infty$. (Osgood.)

12. If $f(u,t)$ is monotone increasing in u for $u > 0$, if $f(0,t) = 0$, if $f_u(u,t)$ decreases as u increases for $u > 0$, and if $-f(-u,t)$ has the same properties as $f(u,t)$, then every solution of $u'' + f(u,t) = 0$ is oscillatory. (Picard.)

13. Consider the equation $u'' + d(t)f(u) = 0$, where $d(t)$ is positive, continuous, monotone increasing, and bounded for $t \geq t_0$, where $f(u)$ is odd and monotone increasing, and where $|f(u_1) - f(u_2)| \leq c_1|u_1 - u_2|$ for $-a < u_1, u_2 < a$, with $a > 0$. Then the particular solution u for which $u = u_1$ and $du/dt = 0$ at $t = t_1$, where $|u_1| < a$ and $f(u_1) = 0$, is oscillatory, and its amplitude decreases monotonically but does not approach zero. (Milne.)

14. If $\phi'(t) > 0$ for $t > t_0$, if $\phi'(t)$ is nonincreasing, and if $\lim\limits_{t \to \infty} \phi(t) = \infty$, then every solution of $u'' + \phi(t)u = 0$ approaches zero as $t \to \infty$, but $\overline{\lim\limits_{t \to \infty}} |u(t) \sqrt{\phi(t)}|$ is positive. (Armellini.)

15. Is the condition that $\phi(t)$ be monotone increasing to ∞ sufficient to guarantee that all solutions of $u'' + \phi(t)u = 0$ approach zero as $t \to \infty$?

16. Consider the equation $u'' + \phi(t)u = 0$, where $\phi(t) \geq 0$ for $a < t < b$. Let u be the solution satisfying the condition $u'(t_0) = 0$, with $a < t_0 < b$ and $u(t_0) = 1$. Then u may be written

$$u = \frac{e^T + e^{-T}}{2}$$

where $T = (t - t_0) \sqrt{\phi(s)}$ and where s is a function of t and where $t_0 \leq s \leq t$. (Petrovitch.)

17. If $\phi(t) < 0$, the solution in the interval between consecutive zeros t_1 and t_2 has the form $u = \cos T$, $T = (t - t_0) \sqrt{-\phi(s)}$, $t_1 \leq s \leq t_2$. Hence $t_1 = t_0 - \pi/2 \sqrt{-\phi(s)}$, $t_2 = t_0 + \pi/2 \sqrt{-\phi(s)}$. (Petrovitch.)

18. If $\phi(t) > 0$, with $t \geq t_0$, the general solution of $u'' - \phi(t)u = 0$ has the form

$$u = c_1 \left[\exp \int_{t_0}^{t} \lambda_1(t_1) \, dt_1 \right] + c_2 \left[\exp \int_{t_0}^{t} \lambda_2(t_1) \, dt_1 \right]$$

where λ_1 and $-\lambda_2$ are nonnegative and bounded if $\phi(t)$ is. (Osgood.)

19. If $\phi(t) > 0$ and is monotone, the amplitudes of the solutions of $u'' + \phi(t)u = 0$ vary monotonically, increasing when $\phi(t)$ is decreasing, decreasing otherwise. Furthermore, if $\phi(t)$ remains finite as $t \to \infty$, the amplitudes remain above a certain bound depending upon $u(0)$. (Murray.)

20. If $\phi(t)$ is monotone and tends to a^2 as $t \to \infty$, then if

$$u'' + \phi(t)u = 0$$

we have

$$\lim_{t \to \infty} \max_{0 \le s \le t} |u| = c_1$$

$$\lim_{t \to \infty} \max_{0 \le s \le t} |u'| = c_2$$

and $c_2 = ac_1$. (Ascoli.)

21. If $\phi(t)$ is nondecreasing, $\max |u|$ is nonincreasing and approaches a finite limit as $t \to \infty$; $\max_{0 \le s \le t} |u'|$ is nondecreasing but may approach ∞ as $t \to \infty$. If $\phi(t)$ is nonincreasing, the above results hold with u and u' interchanged. (Ascoli.)

22. If $\phi'(t) > 0$ and is nondecreasing for $t \ge t_0$ and if $\phi(t + 1/\sqrt{\phi(t)})/\phi(t) \to 1$ as $t \to \infty$, then every solution of $u'' + \phi(t)u = 0$ approaches zero as $t \to \infty$. (Biernacki.)

23. Show that the substitution $u = r \cos \theta$, with $\theta = c \int dt/r^2$, transforms $u'' + \phi(t)u = 0$ into $d^2r/dt^2 - c^2/r^3 + r\phi(t) = 0$.

24. If $\lim_{t \to \infty} \phi(t + c/\sqrt{\phi(t)})/\phi(t) = 1$, $\phi(t) \to \infty$ as $t \to \infty$, then if $\Delta(t)$ is the interval between two successive zeros of $u'' + \phi(t)u = 0$, we have $\lim_{t \to \infty} \Delta(t) \sqrt{\phi(t)}/\pi = 1$. (Wiman.)

25. Under the same hypotheses as in 24, any solution of $u'' - \phi(t)u = 0$ satisfies the relation $\lim_{t \to \infty} u'/u \sqrt{\phi(t)} = +1$ or -1. (Wiman.)

26. Consider the two relations

(a) $\dfrac{d^2u}{dt^2} - p_1(t) \dfrac{du}{dt} - p_2(t)u - q(t) = 0, \ t \ge t_0$

(b) $\dfrac{d^2v}{dt^2} - p_1(t) \dfrac{dv}{dt} - p_2(t)v - q(t) > 0$

where $u(t_0) = u_0 = v(t_0)$ and $u'(t_0) = u_0' = v'(t_0)$. If there exists a solution u of (a) which does not vanish for $t_1 > t > t_0$, then $v > u$ for $t_1 > t > t_0$. (Wilkins.)

27. Let the equation $u'' + p_1(t)u' + p_2(t)u = 0$ possess the property that for $a < t < b$ there exist solutions u_1 and u_2 such that $u_1 > 0$, and $w(u_1, u_2) > 0$, where

$$w = \begin{vmatrix} u_1 & u_1' \\ u_2 & u_2' \end{vmatrix}$$

Then if a function v vanishes for three points in (a,b), there exists an intermediate point s such that $v''(s) + p_1(s)v'(s) + p_2(s)v(s) = 0$. Generalize. (Polya.)

BIBLIOGRAPHY

A more complete account of known results concerning the equation

$$d(k(t)du/dt)/dt + l(t)u = 0$$

will be found in Chap. 3 of the author's survey (1949) mentioned previously.

Section 4

Ascoli, G., *Sul comportamento asintotico degli integrali delle equazioni differenziali del 2° ordine*, Rend. Accad. Lincei, ser. 6, (1935), pp. 234–243.

Bellman, R., *The boundedness of solutions of linear differential equations*, Duke Math. J., vol. 14 (1947), pp. 83–97.

Cacciopoli, R., *Sopra un criterio di stabilita*, Rend. Accad. Lincei, ser. 6 (1930), pp. 251–254.

Wiman, A., *Ueber eine Stabilitätsfrage in der Theorie der linearen Differentialgleichungen*, Acta Math., vol. 66 (1936), pp. 121–145.

Section 5. The first counterexample was given by:

Perron, O., *Ueber ein vermeintliches Stabilitätskriterium*, Nachr. Ges. Wiss., Göttingen, Math.-physik. Kl. Fachgruppe I (1930), pp. 28–29.

The procedure followed above is due to:

Wintner, A., *The adiabatic linear oscillator*, Amer. J. Math., vol. 68 (1946), pp. 385–397. See also:

Ascoli, G., *Osservazioni sopra alcune questioni di stabilita*, Rend. Accad. Lincei, ser. 8, vol. 9 (1950), pp. 210–213.

Section 6

Biernacki, M., *Sur l'équation $x'' + A(t)x = 0$*, Prace Mat. Fiz., vol. 40 (1932), pp. 163–171.

Section 7. The result in the text is a special case of a theorem of Haupt; see:

Haupt, O., *Ueber das asymptotische Verhalten der Lösungen gewisser linearer gewöhnlicher Differentialgleichungen*, Math. Zeit., vol. 48 (1913), pp. 289–292.

For an extensive discussion of the second-order case, see:

Hille, E., *Non-oscillation theorems*, Trans. Amer. Math. Soc., vol. 64 (1948), pp. 234–252.

Section 8

Bellman, R., *A stability property of solutions of linear differential equations*, Duke Math. J., vol. 11 (1944), pp. 513–516.

Section 9

Hardy, G. H., J. E. Littlewood, and G. Polya, *Inequalities*, p. 187, Cambridge University Press, New York, 1934.

See also:

Halperin, I., Ann. of Math., vol. 38 (1937), pp. 889–919, Lemma 2.1.

Landau, E., Math. Ann., vol. 102 (1929), pp. 177–178. Some original results of E. Esclangon are simplified.

Section 10. For the application of the identities of the text to the theory of Sturm-Liouville differential equations, see:

Ince, E. L., *Ordinary differential equations*, Chap. X, London, 1927 (Dover reprint, 1944).

The study of the corresponding properties of general second-order linear differential equations was begun by A. Kneser; see:

Fowler, R. G., *The form near infinity of real continuous solutions of a certain differential equation of the second order*, Quart. J. Math. Oxford Ser., vol. 45 (1914), pp. 289–350.

Kneser, A., *Untersuchungen über die reelen Nullstellen der integrale linearer Differentialgleichungen*, Math. Ann., vol. 42 (1893), pp. 409–435.

Section 11. The result is due to

Liapounoff, A., *Problème général de la stabilité du mouvement*, Ann. Fac. Sci. Univ. Toulouse, ser. 2, vol. 9 (1907), pp. 203–475. Reprinted in the Annals of Mathematics Studies, 1947.

The proof given is due to:

Borg, G., *Ueber die Stabilität gewissen Klassen von linearen Differentialgleichungen*, Ark. Math. Astr. Fys., vol. 31A (1944), no. 1, pp. 460–482.

Section 12. For more detailed results, see:

Hartman, P., *Unrestricted solution fields of almost-separable differential equations*, Trans. Amer. Math. Soc., vol. 63 (1948).

Section 14. For some physical applications of the Liouville transformation, see:

Schelkounoff, S. A., *Solutions of linear and slightly non-linear differential equations*, Quart. Appl. Math., vol. 3 (1945), pp. 349–355.

See also:

Brillouin, L., Quart. Appl. Math., vol. 6 (1948), p. 169; vol. 7 (1949), p. 363.

Section 15. See the paper of Hartman mentioned above and

Bellman, R., *On the asymptotic behavior of solutions of $u'' - (1 + f(t))u = 0$*, Annali Matematica, vol. 31 (1950), pp. 83–91.

Section 18

Kneser, A., *Untersuchung und asymptotische Darstellung der integrale . . .* , J. Reine Angew. Math., vol. 117, 1889, pp. 72–103.

Wintner, A., *Asymptotic integration of the adiabatic oscillator*, Amer. J. Math., vol. 69 (1947), pp. 251–272.

Section 19

Prodi, G., *Nouvi criteri di stabilita per l'equazione $y'' + A(x)y = 0$*, Rend. Accad. Lincei, vol. 10 (1951), pp. 447–451.

CHAPTER 7

THE EMDEN-FOWLER EQUATION

1. Introduction. In this chapter we shall study the important non-linear second-order equation

$$(1) \qquad \frac{d}{dt}\left(t^{\rho} \frac{du}{dt} \right) \pm t^{\sigma} u^{n} = 0$$

This equation has several interesting physical applications, occurring in astrophysics in the form of the Emden equation and in atomic physics in the form of the Fermi-Thomas equation. There seems little doubt that nonlinear equations of this type would enter with greater frequency into mathematical physics, were it more widely known with what ease the properties of the physical solutions can be determined.

Mathematically, the equation possesses great interest: it is a nontrivial, nonlinear differential equation with a large class of solutions whose behavior can be ascertained with astonishing accuracy, despite the fact that the solutions, in general, cannot be obtained explicitly.

In order to isolate this large class of tractable solutions, we employ the concept of *proper solution*, previously encountered in Chap. 5. We recall that a proper solution is one which is real and continuous for $t \geq t_0$.

Henceforth we shall confine ourselves to the consideration of proper solutions alone. In order to remind the reader of this fact, we shall constantly insert this assumption into our hypotheses. This assumption is a natural one as far as physical applications are concerned.

2. Some Preliminary Reductions. We now consider some changes of variable which reduce (1) of Sec. 1 to simpler form.

If $\rho > 1$, set

$$(1) \qquad s = (\rho - 1)^{-1} t^{\rho-1}, \qquad u = (\rho - 1)^{(\rho-\sigma-2)/[(\rho-1)(n-1)]} \frac{v}{s}$$

The equation for v is

$$(2) \qquad \frac{d^2 v}{ds^2} \pm s^{\sigma_1} v^n = 0$$

where

$$(3) \qquad \sigma_1 = \frac{\sigma + \rho}{\rho - 1} - (n + 3)$$

If $\rho < 1$, set

$$
(4) \qquad s = (1 - \rho)^{-1}t^{1-\rho}, \qquad u = (1 - \rho)^{-(\sigma+\rho)/[(n-1)(1-\rho)]}v
$$

Then we have

$$
(5) \qquad \frac{d^2v}{ds^2} \pm s^{\sigma_2}v^n = 0
$$

where

$$
(6) \qquad \sigma_2 = \frac{\sigma + \rho}{1 - \rho}
$$

If $\rho = 1$, set $s = \log t$. The resultant equation is

$$
(7) \qquad \frac{d^2u}{ds^2} \pm e^{(\sigma+1)s}u^n = 0
$$

We begin then by studying the equation

$$
(8) \qquad \frac{d^2u}{dt^2} \pm t^\sigma u^n = 0
$$

For certain values of σ and n, it is possible to reduce (8) to a nonlinear equation with constant coefficients and thereby open the way to the application of the Poincaré-Liapounoff theory to the study of (8).

Let us try a solution of the form $u = ct^w$, c and w being constants. Substituting, we see that there is a solution of this form if

$$
(9) \qquad w = \frac{-(\sigma + 2)}{n - 1}
$$

$$
c = \left[\mp \frac{(\sigma + 2)(\sigma + n + 1)}{(n - 1)^2} \right]^{1/(n-1)}
$$

These equations are meaningless for $n = 1$, in which case the equation is linear and amenable to the method of Chap. 6. We assume henceforth that our equation is actually nonlinear, that is, $n > 1$.

Referring to the value for c given in (9), we see that in general real solutions of this form will exist for the equation $u'' - t^\sigma u^n = 0$ only if $(\sigma + 2)(\sigma + n + 1) > 0$. Whenever these particular solutions exist, we shall see that they are not isolated curiosities but furnish valuable clues to the structure of the set of proper solutions.

Since we are considering only real continuous solutions, the arithmetic nature of n will have considerable influence upon the possible types of proper solutions. It is clear that, in general, because of the presence of the term u^n, proper solutions must be positive. For certain values of n, however, u may take negative values, namely, if $n = p/q$ where q is odd. We see then that for the equations where negative values of u are permissible, either u^n is always positive, or $(-u)^n = -u^n$. These simple observations will explain our apparent prejudice in favor of positive proper solutions.

We will say that n is "odd" if $n = p/q$, with p and q both odd, and that n is "even" if q is even.

It will be necessary to divide the treatment of (8) into subcases, depending upon the sign, and the values of σ and n. As we proceed in this chapter, we shall introduce one device after another up to a certain point, after which these same devices will be used in unison. It is hoped then that, after a certain amount of working and reworking of these sections, the reader will be able to use this same limited number of techniques to handle any equation of similar type that may occur in theory or practice.

3. $u'' - t^\sigma u^n = 0$, $\sigma + n + 1 < 0$, **Positive Proper Solutions.** Our first result is

Theorem 1. *If $\sigma + n + 1 < 0$, all positive proper solutions of*

$$(1) \qquad u'' - t^\sigma u^n = 0$$

have one or the other of the following asymptotic expressions:

$$(2) \qquad u \sim \left[\frac{(\sigma + 2)(\sigma + n + 1)}{(n - 1)^2} \right]^{1/(n-1)} t^{-[(\sigma+2)/(n-1)]}$$

or

$$(3) \qquad u \sim a_1 t + a_2 + [1 + o(1)] \frac{a_2^n t^{\sigma+n+2}}{(\sigma + n + 1)(\sigma + n + 2)}$$

or

$$(4) \qquad u \sim a_2 + [1 + o(1)] \frac{a_2^n t^{\sigma+2}}{(\sigma + 1)(\sigma + 2)}$$

where a_1 and a_2 are arbitrary constants.

Proof. First of all, it is clear that u must be eventually monotone. For if $u' = 0$ at t_0, u can only have a minimum at t_0, since $u'' = t^\sigma u^n > 0$. Hence u is eventually monotone increasing or monotone decreasing. Furthermore u' is monotone increasing, since $u'' > 0$.

Thus there are three cases possible as $t \to \infty$:

$$(5) \qquad \begin{array}{ll} (a) & u' \to 0 \\ (b) & u' \to a \neq 0 \\ (c) & u' \to \infty \end{array}$$

Case (a). If $u' \to 0$, with u' increasing, then $u' < 0$, and u is decreasing. Hence u has a finite limit as $t \to \infty$, since $u > 0$. Moreover this limit is not zero. For if $u'(\infty) = u(\infty) = 0$, we obtain from equation (1)

$$(6) \qquad u'(t) = -\int_t^\infty u'' \, dt = -\int_t^\infty t^\sigma u^n \, dt$$

$$u(t) = -\int_t^\infty u' \, dt = \left(\int_t^\infty \int_t^\infty t_1^\sigma u^n \, dt_1 \right) dt$$

Let $u(t_0) = \delta$ be small. Then since u is monotone decreasing,

$$(7) \qquad \delta = u(t_0) = \int_{t_0}^\infty \left(\int_t^\infty t_1^\sigma u^n \, dt_1 \right) dt \le \delta^n \int_{t_0}^\infty \left(\int_t^\infty t_1^\sigma \, dt_1 \right) dt$$

Since $n > 1$ and since $\sigma + n + 1 < 0$, this last integral converges, and we have a contradiction for δ sufficiently small.

Let then $u(\infty) = a_2 \ne 0$, $u(t) = a_2 + o(1)$ as $t \to \infty$. From (6) we obtain

$$(8) \qquad u'(t) = -a_2^n \int_t^\infty t^\sigma \, dt + o(1) \int_t^\infty t^\sigma \, dt$$

$$= a_2^n \frac{t^{\sigma+1}}{\sigma+1} + o(1)t^{\sigma+1}$$

and thus

$$(9) \qquad u(t) = a_2 - \frac{a_2^n}{\sigma+1} \int_t^\infty t^{\sigma+1} \, dt(1 + o(1))$$

$$= a_2 + \frac{a_2^n t^{\sigma+2}}{(\sigma+1)(\sigma+2)} (1 + o(1))$$

Exercise

1. Using the method of successive approximations, show that a solution of this type exists for $t \ge t_0$, provided that a_2 is suitably chosen.

Case (b).   If $u' \to a_1 \ne 0$, then $u \sim a_1 t$ as $t \to \infty$. Using (6), one then obtains (3).

Exercise

2. Under what conditions does a solution of this type exist?

Case (c).   Referring to (9) of Sec. 2, we see that, if $\sigma + n + 1 < 0$, with $n > 1$, then c is a real constant for all n, so that a particular solution of (1) is $u = ct^w$, where c and w have the values of (9) of Sec. 2. What we wish to show is that this solution is representative of the solutions for which $u' \to \infty$. To do this, we make the substitution

$$(10) \qquad u = ct^w v$$

The equation for v is

$$(11) \qquad t^2 v'' + 2wtv' + w(w-1)(v - v^n) = 0$$

Now perform the further change of variable, $t = e^s$. The resulting equation is

$$(12) \qquad \frac{d^2v}{ds^2} + (2w - 1)\frac{dv}{ds} + w(w - 1)(v - v^n) = 0$$

This equation will play an even more important role below.

Since $u > 0$ and $c > 0$, it follows that $v > 0$. Let us now show that all positive proper solutions of (11) lie in the strip $0 < v < 1$. As soon as v crosses $v = 1$ (it cannot be tangent to $v = 1$ without being identically equal to 1), it must continue increasing monotonically. For if $v' = 0$, we have $v'' = -w(w - 1)(v - v^n) > 0$, and thus only minima can occur. Furthermore any such v cannot approach a finite limit, since we shall show below that if, as $s \to \infty$, $v \to a$, then $a - a^n = 0$, which contradicts $v > 1$ and monotone increasing.

This we prove as follows: Suppose $v \to a$, with $a - a^n \neq 0$. Then from (12), we have, as $s \to \infty$,

$$(13) \qquad \frac{d^2v}{ds^2} + (2w - 1)\frac{dv}{ds} \to c_1 \neq 0$$

Integrating, we obtain, as $s \to \infty$,

$$(14) \qquad \frac{dv}{ds} + (2w - 1)v \sim c_1 s$$

Since $v \to a$, this implies $dv/ds \sim c_1 s$, whence $v \sim c_1 s^2/2$, which contradicts the boundedness of v.

Let us show that $v \to \infty$ is impossible for a proper solution. Making the substitution $dv/ds = p$, (12) becomes

$$(15) \qquad p\frac{dp}{dv} + ap + b(v - v^n) = 0$$

where $a = 2w - 1$, and $b = w(w - 1)$.

Using Theorem 3 of Chap. 5, we see that, as $v \to \infty$, either

$$(16) \qquad p \sim v^k e^{P(v)}$$

where P is a polynomial in v, or

$$(17) \qquad p \sim v^l (\log v)^m$$

If $P(v) \to -\infty$ as $v \to \infty$, then $p \to 0$ and $dp/dv \to 0$. Turning to (15), we see that this results in a contradiction. Hence $P(v) \to \infty$ as $v \to \infty$. This, however, implies $p \geq v^2$ as $v \to \infty$, which is not possible if v is a proper solution, using Lemma 1 of Chap. 5. If $P(v)$ is identically constant, then it follows from (15) that $k = (n + 1)/2 > 1$, and again we have a contradiction. Similarly with (17).

Hence we are left with solutions lying wholly within the strip $0 < v < 1$. These solutions again must be eventually monotone, since any extremum must be a maximum.

Consequently, as $s \to \infty$, $v \to 0$ or $v \to 1$.

The case $v \to 1$ furnishes us with the remaining type of solution of (1), that given by (2), while the solutions corresponding to $v \to 0$ yield the types already found. Incidentally this is another way of showing that $u \to 0$ is impossible, since the rate of decrease of the solutions of the nonlinear equation (11) cannot be too rapid.

Finally let us note that we have made no use of the parity of n.

Exercise

3. Under what conditions do there exist solutions of (12) of the specified type?

---

4. $u'' - t^\sigma u^n = 0$, $\sigma + 2 < 0 < \sigma + n + 1$. In this case we show

Theorem 2. *If $\sigma + 2 < 0 < \sigma + n + 1$, every positive proper solution of*

$$(1) \qquad\qquad u'' - t^\sigma u^n = 0$$

has the asymptotic form

$$(2) \qquad u = a + \frac{a^n t^{\sigma+2}}{(\sigma + 1)(\sigma + 2)} (1 + o(1))$$

Proof. We have again the same three cases:

$$(3) \qquad
\begin{aligned}
&(a) \; u' \to 0 \\
&(b) \; u' \to a \neq 0 \\
&(c) \; u' \to \infty
\end{aligned}$$

as $t \to \infty$.

Let us first show that case (3b) is impossible. If $u' \to a$, then $u \sim at$, and from (1)

$$(4) \qquad\qquad u'' > a_1^n t^{\sigma+n}$$

for $a > a_1 > 0$ and $t \geq t_0$, whence integration yields

$$(5) \qquad\qquad u' > \frac{a_1^n t^{\sigma+n+1}}{\sigma + n + 1} - c_1 \to \infty$$

which is a contradiction. Similarly we show that case (3c) cannot occur. For $u' \to \infty$ implies $u' \geq a$ for large t for some a, and hence $u \geq at$. Reverting to (1), $u'' \geq a^n t^{\sigma+n}$, $u' \geq (a_1^n/\sigma + n + 1)t^{\sigma+n+1} - c_1 \geq t^b$, for some $b > 0$, and thus $u \geq t^{b+1}$ for some $b \geq 0$ as $t \to \infty$. Continuing in this way, we obtain $u \geq t^N$ for every N as $t \to \infty$. Hence from (1), for large t,

$$(6) \qquad u'' > t^\sigma u^n > u^{1+\epsilon}, \qquad \epsilon > 0$$

as $t \to \infty$. If u' is negative, u is decreasing and approaches a finite limit; this is what we wish to show. Hence let us assume u' is positive. If u' is positive, we have from (6)

$$(7) \qquad u'u'' > u^{1+\epsilon}u'$$

which upon integration yields $u'^2 > c_2 u^{2+\epsilon}$ as $t \to \infty$, or $u' > c_3 u^{1+\epsilon/2}$. But this we know is not possible if u is a proper solution.

Consequently we are left with case $(3a)$, where $u' \to 0$. In this case, u' must be negative for large t, since positive u implies, by virtue of the differential equation, that $u'' > 0$, and hence that u' is increasing. Once it has been established that $u' < 0$, it follows that u approaches a limit as $t \to \infty$. The proof given in Sec. 3, (6) and thereafter, shows that this limit is not zero.

Once this point has been settled, we may use the same iteration procedure as before to obtain the asymptotic expression of the solutions.

Exercise

Under what conditions does a solution of the stated type exist?

---

5. $\sigma + 2 < 0$, $\sigma + n + 1 = 0$, $u'' - t^\sigma u^n = 0$. Once more we have the same three cases as $t \to \infty$:

$$(1) \qquad (a)\ u' \to 0$$
$$(b)\ u' \to a \neq 0$$
$$(c)\ u' \to \infty$$

That case $(1b)$ is impossible, follows as above, with (5) of the previous section replaced by

$$(2) \qquad u' > a^n \log t - c_1 \to \infty$$

To rule out $u' \to \infty$ [Case $(1c)$], consider the equation obtained by setting $u = vt$,

$$(3) \qquad t^2 v'' + 2tv' - v^n = 0$$

and then $t = e^s$,

$$(4) \qquad \frac{d^2v}{ds^2} + \frac{dv}{ds} - v^n = 0$$

Since $v > 0$, all solutions are eventually monotone. Using Hardy's theorem as before, we rule out $v \to \infty$. Hence v has a finite limit, which can be only zero, as we see from equation (4). If $v \to 0$, it has the asymptotic expression $v \sim c_1 e^{-s}$, with $c_1 \neq 0$; hence $u = vt \sim c_1$ as $t \to \infty$. That this constant c_1 is not zero may be deduced from (4), as we have just done, or may be shown by virtue of the fact that $\sigma + 2 < 0$, as we have done in the preceding sections.

It is important to keep in mind that there are these alternative approaches to the problem of determining the behavior of the solutions, since in more complicated cases one of the approaches may fail.

Thus we have

Theorem 3. *If $\sigma + 2 < 0 = \sigma + n + 1$, every positive proper solution of*

$$(5) \qquad u'' - t^{\sigma} u^n = 0$$

has the asymptotic form

$$(6) \qquad u = a + \frac{a^n t^{\sigma+2}}{(\sigma + 1)(\sigma + 2)} \, (1 + o(1))$$

as $t \to \infty$.

6. $\sigma + 2 = 0$, $u'' - t^{\sigma} u^n = 0$. We shall prove

Theorem 4. *Every positive proper solution of*

$$(1) \qquad t^2 u'' - u^n = 0$$

has the asymptotic form

$$(2) \qquad u \sim \left(\frac{\log t}{n - 1} \right)^{1/(n-1)}$$

as $t \to \infty$.

Proof. Set $t = e^s$, obtaining

$$(3) \qquad u'' - u' - u^n = 0$$

Since $u > 0$, every solution is eventually monotone; hence u approaches zero, infinity, or a finite limit. The only finite limit it can approach is zero. Using Hardy's theorem as above, we rule out $u \to \infty$ and hence are left with $u \to 0$. To determine the form, set $u' = p$, obtaining

$$(4) \qquad p \frac{dp}{du} - p - u^n = 0$$

and then $u = 1/v$. Discussion of the various possibilities results in (2). We leave the details as an exercise.

Exercise

Under what conditions does a solution of the stated type exist?

---

7. $\sigma + 2 > 0$, $u'' - t^\sigma u^n = 0$

Theorem 5. *If $\sigma + 2 > 0$, every positive proper solution of*

$$(1) \qquad\qquad u'' - t^\sigma u^n = 0$$

has the asymptotic form

$$(2) \qquad\qquad u \sim ct^{-(\sigma+2)/(n-1)}$$

Proof. All solutions are monotone as before. Set $u = ct^w v$, where c and w have the values given in (9) of Sec. 2. The equation for v is

$$(3) \qquad v'' + (2w - 1)v' + w(w - 1)(v - v^n) = 0$$

For σ and n in the range considered, we have

$$(4) \qquad\qquad (2w - 1) < 0 < w(w - 1)$$

Let us now consider the possible alternatives for v; we already have $v > 0$. If v crosses $v = 1$, it must continue monotonically increasing, since any turning point must be a minimum. That v approach a finite limit greater than 1 is impossible, since any finite limit must be a root of $v - v^n = 0$. Hence $v \to \infty$. We now investigate this possibility using Hardy's theorem. Setting $p = v'$, we obtain

$$(5) \qquad\qquad p\frac{dp}{dv} - ap + b(v - v^n) = 0$$

As $v \to \infty$, we must have either

$$(6) \qquad\qquad p \sim e^{P(v)}v^c$$

where P is a polynomial or

$$(7) \qquad\qquad p \sim v^{a_1}(\log v)^{b_2}$$

Evaluation of the constants shows that both cases lead to $p \geq v^{1+\epsilon}$, with $\epsilon > 0$, as $v \to \infty$. Since $p = dv/dt$, this is impossible if we are considering proper solutions.

Hence if $v > 1$, $v \to 1$ as $t \to \infty$, which yields (2).

Now let us consider the solutions in the region $0 < v < 1$. From the ultimate monotonicity of the solutions, $v \to 0$ or $v \to 1$ as $t \to \infty$. We can easily rule out the possibility that $v \to 0$. The characteristic roots of the linear part of (3) are given by $\lambda = -w$, $-(w - 1)$. Since both are positive, it follows that $v = 0$ is a thoroughly instable solution and thus that no other solution of (3) can tend to this as $t \to \infty$. Hence again the only alternative is $v \to 1$, which yields (2).

This concludes the discussion of the positive proper solutions of the equation $u'' - t^\sigma u^n = 0$. Since all proper solutions are monotone, these solutions are ultimately positive or negative. If u is negative, the question arises as to the meaning of u^n. Either n has a value which rules out negative values, or $(-u)^n = \pm u^n$, in which case we can reduce the discussion to the previous case or to the case still to be discussed where $u'' + t^\sigma u^n = 0$.

Exercise

Under what conditions does a solution of the stated type exist?

---

8. $u'' + t^\sigma u^n = 0$, $\sigma + n + 1 < 0$, n **"Odd"**

Theorem 6. *The proper solutions of*

$$(1) \qquad\qquad u'' + t^\sigma u^n = 0$$

possess one of the asymptotic forms

$$(2) \qquad\qquad u = c_7 - \frac{c_6 t^{\sigma+2}}{(\sigma+1)(\sigma+2)}\,(1 + o(1))$$

$$u \sim c_5 t$$

as $t \to \infty$, *if* $\sigma + n + 1 < 0$.

Proof. At this point we introduce a new artifice. We have

$$(3) \qquad\qquad u'u'' + t^\sigma u^n u' = 0$$

whence

$$(4) \qquad\qquad \frac{u'^2}{2} + \int_1^t t^\sigma u^n u'\, dt = c_1$$

Integrating by parts,

$$(5) \qquad\qquad \frac{u'^2}{2} + \frac{t^\sigma u^{n+1}}{n+1} - \sigma \int_1^t t^{\sigma-1} \frac{u^{n+1}}{n+1}\, dt = c_1$$

Since $\sigma < 0$ and since $n + 1$ is even, whence $u^{n+1} \geq 0$, this yields

$$(6) \qquad (a)\ \ u'^2 \leq c_2$$
$$(b)\ \ t^\sigma u^{n+1} \leq c_3$$

Hence

$$(7) \qquad\qquad |u| = O(t^{-\sigma/(n+1)})$$

Returning to our original equation, we obtain

$$(8) \qquad u' + \int_1^t t^\sigma u^n \, dt = c_4$$

The integral is

$$(9) \qquad O \left(\int^t t^\sigma t^{-n\sigma/(n+1)} \, dt \right) = O \left(\int^t t^{\sigma/(n+1)} \, dt \right)$$

Since $\sigma/(n+1) < -1$, the integral converges. Thus $u' \to c_5$ as $t \to \infty$. If $c_5 \neq 0$, $u \sim c_5 t$. If $c_5 = 0$, which is to say, if $u' \to 0$ as $t \to \infty$, we have, in place of (8),

$$(10) \qquad u' = \int_t^\infty t^\sigma u^n \, dt$$

Since $u = O(t^{-\sigma/(n+1)})$, we have

$$(11) \qquad \begin{aligned} u' &= O \left(\int_t^\infty t^\sigma t^{-n\sigma/(n+1)} \, dt \right) = O \left(\int_t^\infty t^{\sigma/(n+1)} \, dt \right) \\ &= O \left(t^{(\sigma+n+1)/(n+1)} \right) \end{aligned}$$

If $\sigma + n + 1 < -(n+1)$, we can conclude that $u \to c_6$ as $t \to \infty$. If not, then $u = O(t^{1-\epsilon})$ for some $\epsilon > 0$, whence, repeating the argument,

$$(12) \qquad \begin{aligned} u' &= O \left(\int_t^\infty t^\sigma u^n \, dt \right) = O \left(\int_t^\infty t^{\sigma+n-n\epsilon} \, dt \right) \\ &= O(t^{\sigma+n+1-n\epsilon}) \end{aligned}$$

and so on, until the exponent is smaller than -1. Thus $u \to c_6$ as $t \to \infty$. Both c_5 and c_6 cannot be zero, as follows from the argument of case $(5a)$ of Sec. 3. The more precise result of (2) may now be obtained by iteration.

Exercise

Under what conditions do solutions of the stated type exist?

---

9. $u'' + t^\sigma u^n = 0$, $\sigma + 2 \geq 0$, n **"Odd."** We now prove the following important result:

Theorem 7. *If $\sigma + 2 \geq 0$, there are no monotone solutions of*

$$(1) \qquad u'' + t^\sigma u^n = 0$$

Proof. Let us consider the case $\sigma + 2 = 0$ first. Assume that there is a monotone increasing solution u, where $u \to l > 0$, with l finite or not, as $t \to \infty$. Setting $t = e^s$, we obtain

$$(2) \qquad v'' - v' + v^n = 0$$

We see that the only finite limit v can have is $v = 0$. To show that $v \to \infty$ is impossible, we use Hardy's theorem as before.

There remains the case $v \to 0$. Setting $p = v'$, we have

$$(3) \qquad p \frac{dp}{dv} - p + v^n = 0$$

We then let $v = 1/u$ and use Hardy's theorem to obtain the possible forms of the solution as $u \to \infty$. In this way we see that no monotone solution exists.

Subsequently we will examine the oscillatory behavior.

Now consider the case $\sigma + 2 > 0$. Take first the case where u is monotone increasing. The limit must be infinite or zero, as we see by considering the equation for $v = u(e^t)$,

$$(4) \qquad v'' - v' + e^{(\sigma+2)t}v^n = 0$$

Let us consider the case $u \to \infty$ first. For $t \geq t_0$, we have from (1)

$$(5) \qquad u'' < -t^\sigma$$

or, integrating,

$$(6) \qquad u' < c_1 - \frac{t^{\sigma+1}}{\sigma + 1} \ (< c_1 - \log t \text{ if } \sigma + 1 = 0)$$

If $\sigma + 1 \geq 0$, this contradicts $u' > 0$ for $t > 0$. If $\sigma + 1 < 0$, integrate (5) between t and ∞, obtaining

$$(7) \qquad u'(t) \geq -\frac{t^{\sigma+1}}{\sigma + 1}$$

Integrating between t_0 and t, we have

$$(8) \qquad u(t) \geq \frac{-t^{\sigma+2}}{(\sigma + 1)(\sigma + 2)} + c_2 \geq t^\epsilon, \qquad \epsilon > 0$$

Returning to the original equation, we have

$$(9) \qquad u'' < -t^{\sigma+n\epsilon}$$

and we repeat this process until $1 + \sigma + n\epsilon > 0$, which will imply $u' < 0$ for t large, and thus a contradiction. The same argument shows that $u \to c_3 > 0$ is impossible, as we already know.

Now consider u monotone decreasing, with $u > 0$. Then $u \to 0$ as $t \to \infty$. Integrating the equation of (1) between t and ∞, we have

$$(10) \qquad u' \Big|_t^\infty + \int_t^\infty t^\sigma u^n \, dt = 0$$
$$u' = \int_t^\infty t^\sigma u^n \, dt > 0$$

which is a contradiction. Since n is odd, the case where $u > 0$ is equivalent to that where $u < 0$.

10. $u'' + t^\sigma u^n = 0$, $\sigma + 2 < 0 < \sigma + n + 1$, $2\sigma + n + 3 < 0$. The arguments used in this and the following section will be more complex and detailed than those given previously. This seems unavoidable, since the set of solutions is actually more varied.

Our first result is

Theorem 8. *If $\sigma + 2 < 0 < \sigma + n + 1$ and if $2\sigma + n + 3 < 0$, with n "odd," all proper solutions of*

$$(1) \qquad u'' + t^\sigma u^n = 0$$

have the asymptotic forms

$$(2) \qquad u \sim ct^w$$

or

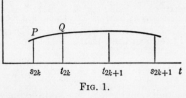

Fig. 1.

$$(3) \qquad u = c_1 - \frac{c_1^n t^{\sigma+2}(1 + o(1))}{(\sigma + 1)(\sigma + 2)}$$

Proof. Let us make the change of variable

$$(4) \qquad u = ct^w v$$

where c and w are determined by the condition that ct^w be a solution of (1). The equation for $v(e^t)$ is then

$$(5) \qquad v'' + (2w - 1)v' + w(w - 1)(v - v^n) = 0$$

where $2w - 1 > 0$ and $w(w - 1) < 0$. Consequently we write this in the form

$$(6) \qquad v'' + av' - b(v - v^n) = 0, \qquad a > 0, b > 0$$

It is here that the condition $2\sigma + n + 3 < 0$ is used in an essential manner.

Multiplying by v' and integrating, we obtain

$$(7) \qquad \frac{v'^2}{2} + a \int_0^t v'^2 \, dt + b\left(\frac{v^{n+1}}{n + 1} - \frac{v^2}{2}\right) = c_1$$

Hence $|v|$ is bounded as $t \to \infty$. From this it follows that

$$(8) \qquad v'^2 < \infty, \qquad \int_0^\infty v'^2 \, dt < \infty$$

From (6) we may then also conclude that $|v''|$ is bounded as $t \to \infty$.

Now let us show that the above conclusions imply that $v' \to 0$ as $t \to \infty$. Let us prove this by contradiction. Let $[t_{2k}, t_{2k+1}]$ (see Fig. 1) be the nth interval, in which $v' \geq a$. Since

$$(9) \qquad a^2 \sum_{n=1}^{\infty} (t_{2k+1} - t_{2k}) \leq \int_0^{\infty} v'^2 \, dt < \infty$$

it follows that $t_{2k+1} - t_{2k} \to 0$. Similarly, consider intervals $[s_{2k}, s_{2k+1}]$, in which $v' \geq a/2$. As above, $s_{2k+1} - s_{2k} \to 0$ as $k \to \infty$.

In the interval $[s_{2k}, t_{2k}]$ the slope of the chord joining PQ is

$$(10) \qquad m = \frac{a - a/2}{t_{2k} - s_{2k}} = v''(\theta), \qquad s_{2k} \leq \theta \leq t_{2k}$$

It follows then that

$$(11) \qquad \varlimsup_{t \to \infty} v''(\theta) = \infty$$

which contradicts the boundedness of $|v''|$. Thus $v' \to 0$ as $t \to \infty$.

Returning to (7), and using $\int_0^{\infty} v'^2 \, dt < \infty$, we obtain

$$(12) \qquad \lim_{t \to \infty} \left(\frac{v^{n+1}}{n+1} - \frac{v^2}{2} \right) = c_3$$

Hence $v \to r$ as $t \to \infty$, where $r^{n+1}/(n+1) - r^2/2 - c_3 = 0$. But r must be a root of $r^n - r = 0$, from (6), and consequently $r = 0$ or 1.

If $r = 1$, we have the desired solution of (2).

Let us then consider the case $r = 0$, where $v \to 0$ and $v' \to 0$ as $t \to \infty$. The linear part of (6), $v'' + av' - bv = 0$, has the associated characteristic roots $-w$, $-w + 1$, the first of which is negative, the second positive. Thus if $v \to 0$ and if $v' \to 0$, $v \sim c_4 e^{-wt}$ as $t \to \infty$. This leads to the solution of (3).

Exercise

Under what conditions do solutions of the above type exist?

---

11. $u'' + t^{\sigma} u^n = 0$, $2\sigma + n + 3 > 0$, $\sigma + 2 < 0 < \sigma + n + 1$. Repeating the preliminary reductions of the previous section, we have the equation

$$(1) \qquad v'' - bv' + c(v^n - v) = 0, \qquad b > 0, c > 0$$

where $b = 2w - 1$ and $c = -w(w - 1)$.

We begin by proving some initial results concerning the behavior of the solutions.

Lemma 1. *There are no proper solutions other than $v(t) = \pm 1$ such that either $v(t) \geq 1$ when $t \geq t_0$ or $v(t) \leq -1$ when $t \geq t_0$. Furthermore there are no solutions such that $v(t)$ approaches 1 from below or -1 from above.*

Proof. Using the above methods, we easily dispose of the possibilities that $v \to \infty$ or $v \to c \neq 1$. Consequently we are left with the case $v \to 1$ from above as $t \to \infty$, if we are considering the case $v \geq 1$. Setting $v = 1 + u$, the resulting equation for u is

$$(2) \qquad u'' - bu' + c[(n - 1)u + O(u^2)] = 0$$

The characteristic roots of the linear equation either are both positive or possess positive real parts. Therefore $u = 0$, that is, $v = 1$, is an unstable solution. This also disposes of the possibility that $v \to 1$ from below as $t \to \infty$. The argument for $v(t) \leq -1$ or $v \to -1$ from above is similar.

Lemma 2. *If* $-1 \leq v \leq 1$, *with* $v \neq \pm 1$, *then*

$$(3) \qquad v \sim c_1 e^{-wt}, \qquad w = -\frac{(\sigma + 2)}{(n - 1)}$$

as $t \to \infty$.

Proof. No oscillatory solutions are possible, as we see by looking at the sign of $v - v^n$. Since v cannot approach $l \neq 0$ or 1, and cannot tend to 1, it must necessarily approach zero as $t \to \infty$. We can then determine the behavior using the theory of Poincaré-Liapounoff or Hardy's theorem, setting $v = 1/u$.

Finally

Lemma 3. *If* $v(t)$ *is not of the above type, or* $+1$, *then as* $t \to \infty$, $v(t)$ *cuts* $+1$ *infinitely often, and* $\overline{\lim} |v| = \infty$.

Proof. Let us assume, without loss of generality, that v intersects $v = 1$ infinitely often. Let $\{t_s\}$ be the sequence of intersections (see Fig. 2).

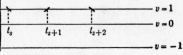

FIG. 2.

Multiplying the equation of (1) by v' and integrating, we obtain

$$(4) \qquad \frac{v'^2}{2} \Big|_{t_s}^{t_{s+1}} - b \int_{t_s}^{t_{s+1}} v'^2 \, dt = 0$$

Hence

$$(5) \qquad \sum_{s=1}^{\infty} [v'^2(t_{s+1}) - v'^2(t_s)] = b \int_0^{\infty} v'^2 \, dt$$

Let us show that $\int_0^{\infty} v'^2 \, dt = \infty$, which will show that $v'^2(t_s) \to \infty$. Assume for the moment that the integral is finite. Then we have from (1), upon multiplication by v' and integration,

$$(6) \qquad v'^2 - b \int_0^t v'^2 \, dt + c \left(\frac{v^{n+1}}{n+1} - \frac{v^2}{2} \right) = c_1$$

This together with $\int_0^\infty v'^2 \, dt < \infty$ would imply that $|v|$ and $|v'|$ are bounded. From the differential equation we have that $|v''|$ is bounded. But we showed in Sec. 10, relations (9) and the following, that these facts imply that $v' \to 0$ as $t \to \infty$. This, however, contradicts (4), which shows that $v'(t_{s+1})^2 > v'(t_s)^2$. Hence we have demonstrated that $|v'(t_s)| \to \infty$ as $s \to \infty$.

Now let us show that the curve can cross $v = 1$, without also crossing $v = 0$, only a finite number of times.   Consider Fig. 3.   At P, $v'(t_s)$ is very

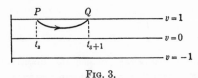

FIG. 3.

large and negative, for t_s large. Since $v''(t_s) = bv'(t_s)$, we see that v' is decreasing at P, whence it is impossible, referring to (1), that $v' = 0$ between t_s and t_{s+1} for $0 < v < 1$. Similarly it follows that the curve $v = v(t)$ must cross $v = -1$ and can return and cross $v = -1$ again only if $|v|$ becomes sufficiently large to counteract the v' term.

This argument then shows that $\varlimsup_{t \to \infty} |v| = +\infty$ and completes the proof of the lemma.

Now that we have obtained this preliminary information, let us attempt to determine the shape of the curve more precisely.

Let us first review what we already have derived. The shape of the curve is as shown in Fig. 4. Furthermore we know that $|v'(t_s)| \to \infty$ as $s \to \infty$, and the argument immediately preceding shows that $|v(\tau_s)| \to \infty$ as $s \to \infty$. We want now to establish the crucial

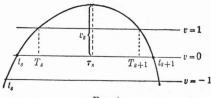

FIG. 4.

Lemma 4. $t_{s+1} - t_s \to 0$ as $s \to \infty$.

Proof. The proof is broken into two parts. We first show that $T_s - t_s \to 0$ and then that $\tau_s - T_s \to 0$ as $s \to \infty$. That $T_s - t_s \to 0$ is an immediate consequence of the fact that $|v'(t_s)| \to \infty$ and of the monotone increasing character of v' in the intervals $[t_s, T_s]$ for large t. This last follows from (1), noting that $0 \leq v \leq 1$ in (t_s, T_s).

Let us now turn to the proof that $\tau_s - T_s \to 0$. Equation (1) may be written

$$(7) \qquad \frac{-d}{dt}(e^{-bt}v')^2 = 2ce^{-2bt}(v^n - v)v'$$

If $t \leq \tau_s$, then $e^{-2b\tau_s} \leq e^{-2bt}$. Integrating (10) over (t, τ_s), we obtain

$$(8) \quad 2ce^{-2b\tau_s}\left(\frac{v_s^{n+1}}{n+1} - \frac{v_s^2}{2} - \frac{v^{n+1}}{n+1} + \frac{v^2}{2}\right) \leq e^{-2bt}v'^2$$
$$\leq 2ce^{-2bt}\left(\frac{v_s^{n+1}}{n+1} - \frac{v_s^2}{2} - \frac{v^{n+1}}{n+1} + \frac{v^2}{2}\right)$$

This yields

$$(9) \quad \sqrt{2c}\, e^{b(t-\tau_s)} \leq v'\left(\frac{v_s^{n+}}{n+1} - \frac{v_s^2}{2} - \frac{v^{n+1}}{n+1} + \frac{v^2}{2}\right)^{-\frac{1}{2}} \leq \sqrt{2c}$$

Integrate between T_s and τ_s, obtaining

$$(10) \quad \frac{\sqrt{2c}}{b}(1 - e^{-b(\tau_s - T_s)}) \leq \int_0^{v_s}\left(\frac{v_s^{n+1}}{n+1} - \frac{v_s^2}{2} - \frac{v^{n+1}}{n+1} + \frac{v^2}{2}\right)^{-\frac{1}{2}} dv$$

The lower limit may be taken to be 0, which, combined with the substitution $v = v_s u$, yields the result

$$(11) \quad \frac{\sqrt{2c}}{b}(1 - e^{-b(\tau_s - T_s)}) \leq v_s^{-[(n-1)/2]}\int_0^1\left(\frac{1 - u^{n+1}}{n+1} - \frac{1 - u^2}{2v_s^{n-1}}\right)^{-\frac{1}{2}} du$$

Since $v_s \to \infty$ as $s \to \infty$, we see that $\tau_s - T_s \to 0$ as $s \to \infty$.

Returning to the u,t plane, and recalling that the t coordinate in the u,t plane is related to the t' coordinate in the v,t' plane by the relation $t = e^{t'}$, we see that if Fig. 5 is the graph of the u curve in the u,t plane, then $t_{s+1}/t_s \to 1$ as $s \to \infty$.

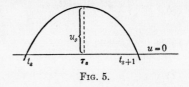

FIG. 5.

Before continuing with the derivation of the asymptotic behavior of the solution, we shall turn to the other σ ranges where oscillatory solutions exist, and show that, in these cases, we also have $t_{s+1}/t_s \to 1$ in the appropriate plane. Once this result has been obtained, a single argument yields the asymptotic behavior in all cases.

12. $u'' + t^\sigma u^n = 0, \sigma + 2 > 0$. We wish to show that in this case too, referring to Fig. 5, we have

Lemma 5. $t_{s+1}/t_s \to 1$ *as* $s \to \infty$.

It is clearly sufficient to prove that $\tau_s/t_s \to 1$, where τ_s is again the point at which u attains its maximum, since a similar argument shows that $t_{s+1}/\tau_s \to 1$, and then Lemma 5 follows. We begin by showing that, if $\sigma > 0$, then

$$(1) \qquad \tau_s^\sigma u_s^{n+1} \geq \tau_{s-1}^\sigma u_{s-1}^{n+1} \geq c_1 > 0$$

From the differential equation we derive

$$(2) \qquad u'^2(t_s) = - \int_{t_s}^{\tau_s} \frac{du'^2}{dt}\, dt = 2 \int_{t_s}^{\tau_s} t^\sigma u^n u'\, dt \leq \frac{2\tau_s^\sigma u_s^{n+1}}{n+1}$$

Similarly, in the interval $[\tau_{s-1}, t_s]$, we have

$$(3) \qquad u'^2(t_s) \geq \frac{2}{n+1} \tau_{s-1}^\sigma u_{s-1}^{n+1}$$

and thus, combining (2) and (3), we obtain (1). Starting with

$$u'u'' + t^\sigma u^n u' = 0$$

we now integrate between t and τ_s and use $t_s \leq t \leq \tau_s$,

$$(4) \qquad \frac{2t^\sigma}{n+1}(u_s^{n+1} - u^{n+1}) \leq u'^2(t) \leq \frac{2t_s^\sigma}{n+1}(u_s^{n+1} - u^{n+1})$$

From this we obtain, integrating between t_s and τ_s,

$$(5) \qquad \sqrt{\frac{2}{n+1}}\, \frac{\tau_s^{(\sigma/2)+1} - t_s^{(\sigma/2)+1}}{(\sigma/2)+1} \leq \int_0^{u_s} \frac{du}{\sqrt{u_s^{n+1} - u^{n+1}}}$$

$$\leq \sqrt{\frac{2}{n+1}}\, t_s^{\sigma/2}(\tau_s - t_s)$$

or

$$(6) \qquad \sqrt{\frac{2}{n+1}}\, \frac{1 - (t_s/\tau_s)^{(\sigma/2)+1}}{(\sigma/2)+1} \leq \frac{1}{\tau_s^{(\sigma/2)+1} u_s^{(n-1)/2}} \int_0^1 \frac{dv}{\sqrt{1 - v^{n+1}}}$$

Since

$$(7) \qquad \tau_s^{\sigma+2} u^{n-1} = [\tau_s^{\frac{(\sigma+2)(n+1)}{n-1}} u_s^{n+1}]^{\frac{n-1}{n+1}}$$

$\tau_s \to \infty$ as $s \to \infty$, and $(\sigma + 2)(n + 1)/(n - 1) > \sigma$, the right-hand side of (6) $\to 0$ as $s \to \infty$, and therefore $t_s/\tau_s \to 1$ as $s \to \infty$.

Let us now consider the case $\sigma < 0, \sigma + 2 > 0$. The same reasoning as used above yields the result

$$(8) \qquad t_s^\sigma u_s^{n+1} \geq t_{s-1}^\sigma u_{s-1}^{n+1} \geq c_1 > 0$$

and its corollary that $u_s \to \infty$ as $s \to \infty$. Since $(\sigma/2) + 1 > 0$, we conclude easily from the above results (4), (5), and (6), appropriately modified to take account of $\sigma < 0$, that again $t_s/\tau_s \to 1$.

13. Asymptotic Behavior of the Oscillatory Solutions of $u'' + t^\sigma u^n = 0$. We shall now show how the preliminary result $t_s/\tau_s \to 1$ may be used to find the asymptotic form of u_s and $t_{s+1} - t_s$. The argument is the same for all cases, and we shall present it in detail for $\sigma > 0$.

From (5) of Sec. 12 we see, using $t_s/\tau_s \to 1$, that

$$(1) \qquad \sqrt{\frac{2}{n+1}} \, (\tau_s - t_s) \sim \left(\frac{\int_0^1 dv / \sqrt{1 - v^{n+1}}}{\tau_s^{\sigma/2} u_s^{(n-1)/2}} \right)$$

In place of this estimate for the length of the total interval $\tau_n - t_n$, which we shall use below, we want first an estimate of $t - t_n$, where $t_n \leq t \leq \tau_n$. Since we have from (4) of Sec. 12

$$(2) \qquad \sqrt{\frac{2}{n+1}} \, \tau_s^{\sigma/2} \sim \frac{u'}{\sqrt{u_s^{n+1} - u^{n+1}}}$$

integration between t_s and $t = t_s + h$ yields

$$(3) \qquad h \sqrt{\frac{2}{n+1}} \, \tau_s^{\sigma/2} \sim \int_0^u \frac{du}{\sqrt{u_s^{n+1} - u^{n+1}}}$$

Returning to the equation, $-d(u'^2)/dt = 2t^\sigma u^n u'$, set $t = t_s + h$ and expand t^σ about t_s, obtaining

$$(4) \qquad -\frac{d}{dt} (u'^2) = 2[t_s^\sigma + h\sigma t_s^{\sigma-1}(1 + \epsilon)]u^n u'$$

where $\epsilon = \epsilon(t) \to 0$ as $t \to \infty$. Now integrate between t_s and τ_s, obtaining

$$(5) \qquad u'(t_s)^2 = \frac{2t_s^\sigma u_s^{n+1}}{n+1} + 2\sigma t_s^{\sigma-1}(1 + \epsilon) \int_{t_s}^{\tau_s} h u^n u' \, dt$$

Employing the estimate for h from (3), this becomes

$$(6) \qquad u'(t_s)^2 = \frac{2t_s^\sigma u_s^{n+1}}{n+1} + \frac{2\sigma t_s^{\sigma-1}(1 + \epsilon)}{\sqrt{2/(n+1)}\tau_s^{\sigma/2}} \int_{t_s}^{\tau_s} u^n u' \left(\int_0^u \frac{dv}{\sqrt{u_s^{n+1} - v^{n+1}}} \right) dt$$

$$= \frac{2t_s^\sigma u_s^{n+1}}{n+1} + \frac{2\sigma t_s^{\sigma-1}(1 + \epsilon)}{\sqrt{2/(n+1)}\tau_s^{\sigma/2}} \frac{u_s^{(n+3)/2}}{n+1} K_1$$

where

$$(7) \qquad K_i = \int_0^1 \sqrt{1 - v^{n+1}} \, dv = \frac{1}{n+1} \frac{\Gamma(3\!/\!2)\Gamma[1/(n+1)]}{\Gamma[3\!/\!2 + (1/n+1)]}$$

The evaluation of the integral in (6) is performed by first changing the independent variable from t to u and then letting $u = u_s v$.

We now have the equality

$$(8) \qquad u'(t_s)^2 = \frac{2t_s^\sigma u_s^{n+1}}{n+1} \left[1 + \frac{\sigma(1+\epsilon)K_1}{\sqrt{2/n+1}\ t_s \tau_s^{\sigma/2} u_s^{(n-1)/2}} \right]$$

Similarly, from the other half of the wave, we obtain

$$(9) \qquad u'(t_s)^2 = \frac{2}{n+1}\ t_s^\sigma u_{s-1}^{n+1} \left[1 - \frac{\sigma(1+\epsilon)K_1}{\sqrt{2/n+1}\ t_s \tau_{s-1}^{\sigma/2} u_{s-1}^{(n-1)/2}} \right]$$

Returning to equation (1) and the estimate for $\tau_s - t_s$, we may simplify (8) and (9) as follows:

$$(10) \qquad \begin{aligned} u'(t_s)^2 &= \frac{2}{n+1}\ t_s^\sigma u_s^{n+1} \left[1 + \frac{\sigma(1+\epsilon)K_1}{t_s K_2}\ (\tau_s - t_s) \right] \\ &= \frac{2t_s^\sigma}{n+1}\ u_{s-1}^{n+1} \left[1 - \frac{(1+\epsilon)K_1}{t_s K_2}\ (t_s - \tau_{s-1}) \right] \end{aligned}$$

where

$$(11) \qquad K_2 = \int_0^1 \frac{dv}{\sqrt{1-v^{n+1}}} = \frac{1}{n+1} \frac{\Gamma(\tfrac{1}{2})\Gamma[1/(n+1)]}{\Gamma[\tfrac{1}{2}+(1/n+1)]}$$

Since $(\tau_s - t_s)/\tau_s$ is small, as is also $(t_s - \tau_{s-1})/t_s$, we write these relations in (10) in the form

$$(12) \qquad \begin{aligned} u'(t_s)^2 &= \frac{2t_s^\sigma u_s^{n+1}}{n+1} \left(1 + \frac{\tau_s - t_s}{t_s} \right)^{\sigma K_1(1+\epsilon)/K_2} \\ &= \frac{2t_s^\sigma u_s^{n+1}}{n+1} \left(\frac{\tau_s}{t_s} \right)^{\sigma K_1(1+\epsilon)/K_2} \end{aligned}$$

Similarly,

$$(13) \qquad u'(t_s)^2 = \frac{2t_s^\sigma u_{s-1}^{n+1}}{n+1} \left[\frac{\tau_{s-1}}{t_s} \right]^{\sigma K_1(1+\epsilon)/K_2}$$

Hence

$$(14) \qquad 1 = \left(\frac{u_s}{u_{s-1}} \right)^{n+1} \left(\frac{\tau_s}{\tau_{s-1}} \right)^{\sigma K_1(1+\epsilon_1)/K_2}$$

Multiplying over s, we obtain

$$(15) \qquad u_s = \tau_s^{-\sigma K_1(1+\epsilon)/K_2} \prod_s \left(\frac{\tau_{s-1}}{\tau_s} \right)^{\epsilon_s} = \tau_s^{-\sigma K_1/[K_2(n+1)]+\epsilon_2}$$

where $\epsilon_2 \to 0$ as $t \to \infty$. Replacing K_1 and K_2 by their numerical values,

$$(16) \qquad u_s = \tau_s^{(-\sigma/n+3)+\epsilon}$$

From this, we obtain, returning to (1),

(17) $$t_{s+1} - t_s = t_s^{(-2\sigma/n+3)+\epsilon}$$

A further use of the above methods yields the precise results,

(18) $$u_s \sim c_1 \tau_s^{-\sigma/(n+3)}$$

$$t_{s+1} - t_s \sim \frac{1}{c_1^{(n-1)/2}} \sqrt{\frac{2}{n+1}} \frac{\Gamma(\frac{1}{2})\Gamma(1/n+1)}{\Gamma[(\frac{1}{2}) + (1/n+1)]} \tau_s^{-2\sigma/(n+3)}$$

where c_1 is a constant.

The same results hold over the other σ ranges, and we leave their derivations as exercises.

14. The Equations $u'' \pm e^{\lambda t} u^n = 0$. Since the methods we have applied in the previous sections are equally applicable to the equations $u'' + e^{\lambda t} u^n = 0$, we shall state the results and leave the proofs as exercises.

Consider first the equation $u'' + e^{\lambda t} u^n = 0$, and define

(1) $$w = -\frac{\lambda}{n-1} \qquad c = w^{2/(n-1)}$$

Exercises

1. If $\lambda > 0$ and $n = p/q$, where p and q are both odd, then all proper solutions are oscillatory.

2. If $\lambda > 0$ and $n = p/q$, where p is even and q is odd, then the proper solutions form a one-dimensional manifold, all asymptotic to $-ce^{wt}$.

3. If n is irrational, or rational with even denominator, there are no proper solutions.

4. If $\lambda < 0$ and $n = p/q$, a rational number not of the form even/odd, the proper solutions form a two-dimensional manifold with the parameters $a_1 = \lim_{t \to \infty} u'$, $a_0 = \lim_{t \to \infty} (u - a_1 t)$.

5. Show that, under the conditions of Exercise 4,

$$u = a_1 t + a_0 + \left(\int_t^\infty (t - s) e^{\lambda s} (a_1 s)^n \, ds \right) (1 + o(1))$$

6. What restrictions must be imposed upon a_0 and a_1 if we wish solutions which exist for $0 < t < \infty$?

7. If $\lambda < 0$ and $n = p/q$, where p is even and q is odd, show that, in addition to the above solutions, there is a one-dimensional manifold of proper solutions asymptotic to $-ce^{wt}$.

In the following two exercises we consider the equation $u'' - e^{\lambda t} u^n = 0$.

8. Show that, if $\lambda > 0$, the proper solutions form a one-dimensional manifold, asymptotic to ce^{wt}.

9. Show that, if $\lambda < 0$, in addition to the above one-dimensional manifold, there exists a two-dimensional manifold of solutions with parameters a_0 and a_1 as above.   If $a_1 \neq 0$, we have

$$u = a_1 t + a_0 + \left(\int_t^\infty (t - s) e^{\lambda s} (a_1 s)^n \, ds \right) (1 + o(1))$$

What conditions must be imposed upon a_0 and a_1 if we want solutions to exist for all $t > 0$?

BIBLIOGRAPHY

For the physical origin of the Emden-Fowler equation, see:

Emden, R., *Gaskugeln, Anwendungen der mechanischen Warmentheorie auf Kosmologie und meteorologische Probleme*, Chap. XII, B. G. Teubner, Leipzig, 1907.

The preceding chapter is a slightly simplified and unified presentation of the contents of

Fowler, R. H., *The form near infinity of real, continuous solutions of a certain differential equation of the second order*, Quart. J. Math., vol. 45 (1914), pp. 289–350.

———, *The solution of Emden's and similar equations*, Monthly Notices of the Royal Astr. Soc., vol. 91 (1920), pp. 63–91.

———, *Further studies of Emden's and similar differential equations*, Quart. J. Math., vol. 2 (1931), pp. 259–288.

For a geometric treatment of some of the above cases, see

Hopf, E., Monthly Notices of the Royal Astr. Soc., vol. 91 (1920), p. 653.

A later paper on the Emden equation is:

Sansone, G., *Sulle soluzioni di Emden della equazione di Fowler*, Univ. Roma e. 1st Naz. Alta. Mat. Rend. Mat. e. Appl., ser. 5, vol. 1 (1940), pp. 163–176.

INDEX

SOME DOVER SCIENCE BOOKS

SOME DOVER SCIENCE BOOKS

WHAT IS SCIENCE?,
Norman Campbell
This excellent introduction explains scientific method, role of mathematics,
types of scientific laws. Contents: 2 aspects of science, science & nature, laws of
science, discovery of laws, explanation of laws, measurement & numerical laws,
applications of science. 192pp. 5⅜ x 8.   Paperbound $1.25

FADS AND FALLACIES IN THE NAME OF SCIENCE,
Martin Gardner
Examines various cults, quack systems, frauds, delusions which at various times
have masqueraded as science. Accounts of hollow-earth fanatics like Symmes;
Velikovsky and wandering planets; Hoerbiger; Bellamy and the theory of
multiple moons; Charles Fort; dowsing, pseudoscientific methods for finding
water, ores, oil. Sections on naturopathy, iridiagnosis, zone therapy, food fads,
etc. Analytical accounts of Wilhelm Reich and orgone sex energy; L. Ron
Hubbard and Dianetics; A. Korzybski and General Semantics; many others.
Brought up to date to include Bridey Murphy, others. Not just a collection of
anecdotes, but a fair, reasoned appraisal of eccentric theory. Formerly titled
In the Name of Science. Preface. Index. x + 384pp. 5⅜ x 8.
                                                            Paperbound $1.85

PHYSICS, THE PIONEER SCIENCE,
L. W. Taylor
First thorough text to place all important physical phenomena in cultural-
historical framework; remains best work of its kind. Exposition of physical
laws, theories developed chronologically, with great historical, illustrative
experiments diagrammed, described, worked out mathematically. Excellent
physics text for self-study as well as class work. Vol. 1: Heat, Sound: motion,
acceleration, gravitation, conservation of energy, heat engines, rotation, heat,
mechanical energy, etc. 211 illus. 407pp. 5⅜ x 8. Vol. 2: Light, Electricity:
images, lenses, prisms, magnetism, Ohm's law, dynamos, telegraph, quantum
theory, decline of mechanical view of nature, etc. Bibliography. 13 table
appendix. Index. 551 illus. 2 color plates. 508pp. 5⅜ x 8.
                                  Vol. 1 Paperbound $2.25, Vol. 2 Paperbound $2.25,
                                                                The set $4.50

THE EVOLUTION OF SCIENTIFIC THOUGHT FROM NEWTON TO EINSTEIN,
A. d'Abro
Einstein's special and general theories of relativity, with their historical implica-
tions, are analyzed in non-technical terms. Excellent accounts of the contributions of Newton, Riemann, Weyl, Planck, Eddington, Maxwell, Lorentz and
others are treated in terms of space and time, equations of electromagnetics,
finiteness of the universe, methodology of science. 21 diagrams. 482pp. 5⅜ x 8.
                                                            Paperbound $2.50

CHANCE, LUCK AND STATISTICS: THE SCIENCE OF CHANCE,
Horace C. Levinson

Theory of probability and science of statistics in simple, non-technical language. Part I deals with theory of probability, covering odd superstitions in regard to "luck," the meaning of betting odds, the law of mathematical expectation, gambling, and applications in poker, roulette, lotteries, dice, bridge, and other games of chance. Part II discusses the misuse of statistics, the concept of statistical probabilities, normal and skew frequency distributions, and statistics applied to various fields—birth rates, stock speculation, insurance rates, advertising, etc. "Presented in an easy humorous style which I consider the best kind of expository writing," Prof. A. C. Cohen, Industry Quality Control. Enlarged revised edition. Formerly titled *The Science of Chance*. Preface and two new appendices by the author. Index. xiv + 365pp. 5⅜ x 8. Paperbound $2.00

BASIC ELECTRONICS,
prepared by the U.S. Navy Training Publications Center

A thorough and comprehensive manual on the fundamentals of electronics. Written clearly, it is equally useful for self-study or course work for those with a knowledge of the principles of basic electricity. Partial contents: Operating Principles of the Electron Tube; Introduction to Transistors; Power Supplies for Electronic Equipment; Tuned Circuits; Electron-Tube Amplifiers; Audio Power Amplifiers; Oscillators; Transmitters; Transmission Lines; Antennas and Propagation; Introduction to Computers; and related topics. Appendix. Index. Hundreds of illustrations and diagrams. vi + 471pp. 6½ x 9¼.

Paperbound $2.75

BASIC THEORY AND APPLICATION OF TRANSISTORS,
prepared by the U.S. Department of the Army

An introductory manual prepared for an army training program. One of the finest available surveys of theory and application of transistor design and operation. Minimal knowledge of physics and theory of electron tubes required. Suitable for textbook use, course supplement, or home study. Chapters: Introduction; fundamental theory of transistors; transistor amplifier fundamentals; parameters, equivalent circuits, and characteristic curves; bias stabilization; transistor analysis and comparison using characteristic curves and charts; audio amplifiers; tuned amplifiers; wide-band amplifiers; oscillators; pulse and switching circuits; modulation, mixing, and demodulation; and additional semiconductor devices. Unabridged, corrected edition. 240 schematic drawings, photographs, wiring diagrams, etc. 2 Appendices. Glossary. Index. 263pp. 6½ x 9¼.

Paperbound $1.25

GUIDE TO THE LITERATURE OF MATHEMATICS AND PHYSICS,
N. G. Parke III

Over 5000 entries included under approximately 120 major subject headings of selected most important books, monographs, periodicals, articles in English, plus important works in German, French, Italian, Spanish, Russian (many recently available works). Covers every branch of physics, math, related engineering. Includes author, title, edition, publisher, place, date, number of volumes, number of pages. A 40-page introduction on the basic problems of research and study provides useful information on the organization and use of libraries, the psychology of learning, etc. This reference work will save you hours of time. 2nd revised edition. Indices of authors, subjects, 464pp. 5⅜ x 8.

Paperbound $2.75

THE RISE OF THE NEW PHYSICS (formerly THE DECLINE OF MECHANISM),
A. d'Abro
This authoritative and comprehensive 2-volume exposition is unique in scientific publishing. Written for intelligent readers not familiar with higher mathematics, it is the only thorough explanation in non-technical language of modern mathematical-physical theory. Combining both history and exposition, it ranges from classical Newtonian concepts up through the electronic theories of Dirac and Heisenberg, the statistical mechanics of Fermi, and Einstein's relativity theories. "A must for anyone doing serious study in the physical sciences," J. of Franklin Inst. 97 illustrations. 991pp. 2 volumes.

T3, T4 Two volume set, paperbound $5.50

THE STRANGE STORY OF THE QUANTUM, AN ACCOUNT FOR THE GENERAL READER OF THE GROWTH OF IDEAS UNDERLYING OUR PRESENT ATOMIC KNOWLEDGE, B. Hoffmann
Presents lucidly and expertly, with barest amount of mathematics, the problems and theories which led to modern quantum physics. Dr. Hoffmann begins with the closing years of the 19th century, when certain trifling discrepancies were noticed, and with illuminating analogies and examples takes you through the brilliant concepts of Planck, Einstein, Pauli, de Broglie, Bohr, Schroedinger, Heisenberg, Dirac, Sommerfeld, Feynman, etc. This edition includes a new, long postscript carrying the story through 1958. "Of the books attempting an account of the history and contents of our modern atomic physics which have come to my attention, this is the best," H. Margenau, Yale University, in American Journal of Physics. 32 tables and line illustrations. Index. 275pp. 5⅜ x 8.

T518 Paperbound $2.00

GREAT IDEAS AND THEORIES OF MODERN COSMOLOGY,
Jagjit Singh
The theories of Jeans, Eddington, Milne, Kant, Bondi, Gold, Newton, Einstein, Gamow, Hoyle, Dirac, Kuiper, Hubble, Weizsäcker and many others on such cosmological questions as the origin of the universe, space and time, planet formation, "continuous creation," the birth, life, and death of the stars, the origin of the galaxies, etc. By the author of the popular Great Ideas of Modern Mathematics. A gifted popularizer of science, he makes the most difficult abstractions crystal-clear even to the most non-mathematical reader. Index. xii + 276pp. 5⅜ x 8½

T925 Paperbound $2.00

GREAT IDEAS OF MODERN MATHEMATICS: THEIR NATURE AND USE,
Jagjit Singh
Reader with only high school math will understand main mathematical ideas of modern physics, astronomy, genetics, psychology, evolution, etc., better than many who use them as tools, but comprehend little of their basic structure. Author uses his wide knowledge of non-mathematical fields in brilliant exposition of differential equations, matrices, group theory, logic, statistics, problems of mathematical foundations, imaginary numbers, vectors, etc. Original publications, appendices. indexes. 65 illustr. 322pp. 5⅜ x 8. T587 Paperbound $2.00

THE MATHEMATICS OF GREAT AMATEURS, Julian L. Coolidge
Great discoveries made by poets, theologians, philosophers, artists and other non-mathematicians: Omar Khayyam, Leonardo da Vinci, Albrecht Dürer, John Napier, Pascal, Diderot, Bolzano, etc. Surprising accounts of what can result from a non-professional preoccupation with the oldest of sciences. 56 figures. viii + 211pp. 5⅜ x 8½.

S1009 Paperbound $2.00

COLLEGE ALGEBRA, *H. B. Fine*

Standard college text that gives a systematic and deductive structure to algebra; comprehensive, connected, with emphasis on theory. Discusses the commutative, associative, and distributive laws of number in unusual detail, and goes on with undetermined coefficients, quadratic equations, progressions, logarithms, permutations, probability, power series, and much more. Still most valuable elementary-intermediate text on the science and structure of algebra. Index. 1560 problems, all with answers. x + 631pp. 5⅜ x 8.   Paperbound $2.75

HIGHER MATHEMATICS FOR STUDENTS OF CHEMISTRY AND PHYSICS, *J. W. Mellor*

Not abstract, but practical, building its problems out of familiar laboratory material, this covers differential calculus, coordinate, analytical geometry, functions, integral calculus, infinite series, numerical equations, differential equations, Fourier's theorem, probability, theory of errors, calculus of variations, determinants. "If the reader is not familiar with this book, it will repay him to examine it," *Chem. & Engineering News.* 800 problems. 189 figures. Bibliography. xxi + 641pp. 5⅜ x 8.   Paperbound $2.50

TRIGONOMETRY REFRESHER FOR TECHNICAL MEN, *A. A. Klaf*

A modern question and answer text on plane and spherical trigonometry. Part I covers plane trigonometry: angles, quadrants, trigonometrical functions, graphical representation, interpolation, equations, logarithms, solution of triangles, slide rules, etc. Part II discusses applications to navigation, surveying, elasticity, architecture, and engineering. Small angles, periodic functions, vectors, polar coordinates, De Moivre's theorem, fully covered. Part III is devoted to spherical trigonometry and the solution of spherical triangles, with applications to terrestrial and astronomical problems. Special time-savers for numerical calculation. 913 questions answered for you! 1738 problems; answers to odd numbers. 494 figures. 14 pages of functions, formulae. Index. x + 629pp. 5⅜ x 8.
     Paperbound $2.00

CALCULUS REFRESHER FOR TECHNICAL MEN, *A. A. Klaf*

Not an ordinary textbook but a unique refresher for engineers, technicians, and students. An examination of the most important aspects of differential and integral calculus by means of 756 key questions. Part I covers simple differential calculus: constants, variables, functions, increments, derivatives, logarithms, curvature, etc. Part II treats fundamental concepts of integration: inspection, substitution, transformation, reduction, areas and volumes, mean value, successive and partial integration, double and triple integration. Stresses practical aspects! A 50 page section gives applications to civil and nautical engineering, electricity, stress and strain, elasticity, industrial engineering, and similar fields. 756 questions answered. 556 problems; solutions to odd numbers. 36 pages of constants, formulae. Index. v + 431pp. 5⅜ x 8.   Paperbound $2.00

INTRODUCTION TO THE THEORY OF GROUPS OF FINITE ORDER, *R. Carmichael*

Examines fundamental theorems and their application. Beginning with sets, systems, permutations, etc., it progresses in easy stages through important types of groups: Abelian, prime power, permutation, etc. Except 1 chapter where matrices are desirable, no higher math needed. 783 exercises, problems. Index. xvi + 447pp. 5⅜ x 8.   Paperbound $3.00

CATALOGUE OF DOVER BOOKS

FIVE VOLUME "THEORY OF FUNCTIONS" SET BY KONRAD KNOPP

This five-volume set, prepared by Konrad Knopp, provides a complete and readily followed account of theory of functions. Proofs are given concisely, yet without sacrifice of completeness or rigor. These volumes are used as texts by such universities as M.I.T., University of Chicago, N. Y. City College, and many others. "Excellent introduction . . . remarkably readable, concise, clear, rigorous," *Journal of the American Statistical Association.*

ELEMENTS OF THE THEORY OF FUNCTIONS,
Konrad Knopp
This book provides the student with background for further volumes in this set, or texts on a similar level. Partial contents: foundations, system of complex numbers and the Gaussian plane of numbers, Riemann sphere of numbers, mapping by linear functions, normal forms, the logarithm, the cyclometric functions and binomial series. "Not only for the young student, but also for the student who knows all about what is in it," *Mathematical Journal.* Bibliography. Index. 140pp. 5⅜ x 8. Paperbound $1.50

THEORY OF FUNCTIONS, PART I,
Konrad Knopp
With volume II, this book provides coverage of basic concepts and theorems. Partial contents: numbers and points, functions of a complex variable, integral of a continuous function, Cauchy's integral theorem, Cauchy's integral formulae, series with variable terms, expansion of analytic functions in power series, analytic continuation and complete definition of analytic functions, entire transcendental functions, Laurent expansion, types of singularities. Bibliography. Index. vii + 146pp. 5⅜ x 8. Paperbound $1.35

THEORY OF FUNCTIONS, PART II,
Konrad Knopp
Application and further development of general theory, special topics. Single valued functions. Entire, Weierstrass, Meromorphic functions. Riemann surfaces. Algebraic functions. Analytical configuration, Riemann surface. Bibliography. Index. x + 150pp. 5⅜ x 8. Paperbound $1.35

PROBLEM BOOK IN THE THEORY OF FUNCTIONS, VOLUME 1.
Konrad Knopp
Problems in elementary theory, for use with Knopp's *Theory of Functions,* or any other text, arranged according to increasing difficulty. Fundamental concepts, sequences of numbers and infinite series, complex variable, integral theorems, development in series, conformal mapping. 182 problems. Answers. viii + 126pp. 5⅜ x 8. Paperbound $1.35

PROBLEM BOOK IN THE THEORY OF FUNCTIONS, VOLUME 2,
Konrad Knopp
Advanced theory of functions, to be used either with Knopp's *Theory of Functions,* or any other comparable text. Singularities, entire & meromorphic functions, periodic, analytic, continuation, multiple-valued functions, Riemann surfaces, conformal mapping. Includes a section of additional elementary problems. "The difficult task of selecting from the immense material of the modern theory of functions the problems just within the reach of the beginner is here masterfully accomplished," *Am. Math. Soc.* Answers. 138pp. 5⅜ x 8. Paperbound $1.50

APPLIED OPTICS AND OPTICAL DESIGN,
A. E. Conrady

With publication of vol. 2, standard work for designers in optics is now complete for first time. Only work of its kind in English; only detailed work for practical designer and self-taught. Requires, for bulk of work, no math above trig. Step-by-step exposition, from fundamental concepts of geometrical, physical optics, to systematic study, design, of almost all types of optical systems. Vol. 1: all ordinary ray-tracing methods; primary aberrations; necessary higher aberration for design of telescopes, low-power microscopes, photographic equipment. Vol. 2: (Completed from author's notes by R. Kingslake, Dir. Optical Design, Eastman Kodak.) Special attention to high-power microscope, anastigmatic photographic objectives. "An indispensable work," *J., Optical Soc. of Amer.* Index. Bibliography. 193 diagrams. 852pp. 6⅛ x 9¼.

Two volume set, paperbound $7.00

MECHANICS OF THE GYROSCOPE, THE DYNAMICS OF ROTATION,
R. F. Deimel, Professor of Mechanical Engineering at Stevens Institute of Technology

Elementary general treatment of dynamics of rotation, with special application of gyroscopic phenomena. No knowledge of vectors needed. Velocity of a moving curve, acceleration to a point, general equations of motion, gyroscopic horizon, free gyro, motion of discs, the damped gyro, 103 similar topics. Exercises. 75 figures. 208pp. 5⅜ x 8.

Paperbound $1.75

STRENGTH OF MATERIALS,
J. P. Den Hartog

Full, clear treatment of elementary material (tension, torsion, bending, compound stresses, deflection of beams, etc.), plus much advanced material on engineering methods of great practical value: full treatment of the Mohr circle, lucid elementary discussions of the theory of the center of shear and the "Myosotis" method of calculating beam deflections, reinforced concrete, plastic deformations, photoelasticity, etc. In all sections, both general principles and concrete applications are given. Index. 186 figures (160 others in problem section). 350 problems, all with answers. List of formulas. viii + 323pp. 5⅜ x 8.

Paperbound $2.00

HYDRAULIC TRANSIENTS,
G. R. Rich

The best text in hydraulics ever printed in English . . . by former Chief Design Engineer for T.V.A. Provides a transition from the basic differential equations of hydraulic transient theory to the arithmetic integration computation required by practicing engineers. Sections cover Water Hammer, Turbine Speed Regulation, Stability of Governing, Water-Hammer Pressures in Pump Discharge Lines, The Differential and Restricted Orifice Surge Tanks, The Normalized Surge Tank Charts of Calame and Gaden, Navigation Locks, Surges in Power Canals—Tidal Harmonics, etc. Revised and enlarged. Author's prefaces. Index. xiv + 409pp. 5⅜ x 8½.

Paperbound $2.50

Prices subject to change without notice.

Available at your book dealer or write for free catalogue to Dept. Adsci, Dover Publications, Inc., 180 Varick St., N.Y., N.Y. 10014. Dover publishes more than 150 books each year on science, elementary and advanced mathematics, biology, music, art, literary history, social sciences and other areas.